WONDER CHEMISTRY
HOW CAN YOU RELATE MODERN CHEMISTRY IN YOUR DAY TO DAY LIFE EXPERIENCES?

WONDER CHEMISTRY
HOW CAN YOU RELATE MODERN CHEMISTRY IN YOUR DAY TO DAY LIFE EXPERIENCES?

Aditi Kwatra

Publishers Name
Delhi

Copyright © 2019 by Aditi Kwatra
All rights reserved. This book or any portion thereof may not be reproduced or used in any manner whatsoever without the express written permission of the publisher except for the use of brief quotations in a book review.

ISBN : 978-1-64783-821-8

Typesetting & Designing
JEE-VEE Graphics Delhi # 9911902109

Price: 399.00

Printed by
India

Dedication

I dedicate this book to my Gurus and Parents Mr. MR Khanna, Mrs. Vijay Laxmi and Mr. Ramesh Kwatra, Mrs Dinesh Kwatra, my Husband Mr. Bharat Kwatra, my loving son Vedant Kwatra and all my dear family members: Mr. Anchit and Mrs. Neha Bagga; Mr. Darpan and Mrs. Bhavika Kwatra; Mr.Karan and Mrs. Tanya Khanna and to my beloved friends.

Without their patience, understanding, support and most above all love, the completion of this work would not have been possible.

Special thanks are due to Mr. Manuj Bajaj without whom this project could not have accomplished and to Mr. Arpit Jain and Mr. Lakshay Kwatra for taking pains in publishing the book in time.

I would like to express my gratitude to many people who saw me through this book; to all those who provided support, talked things over, read, wrote, offered comments, allowed me to quote their remarks and assisted in the editing, proofreading and design.

Preface

This book relates Chemistry to different aspects of our daily lives. Research in chemistry has led to life changing discoveries, allowed us to progress as a society, has made our lives safer and even allowed us to live longer. Without Chemistry, we would not have nearly all of the products that we wear, eat and use daily. It not only has improved our daily lives, but has made a significant impact on how our society has evolved and flourished.

The purpose of this book is to widen our knowledge about the utility of chemistry in our daily life. Since Chemistry has a vast role in different aspects of our lives, there is a need for us to recognize the importance of Chemistry in the real world around us. I am sure that this book would generate curiosity among the readers to know more about chemical sciences and in knowing about benefits and hazards of different chemicals around us.

Please note that some terminologies have been super-scripted in the book. They have been explained in the Glossary at the end.

Contents

Preface	7
Chapter 1: Introduction	15
Contribution of Chemistry in Improving our daily life	15
Chapter 2: Chemistry in Medicines and Health Care	19
Introduction	19
Antacids	23
Analgesics	23
Antibiotics	25
When To Take Them?	25
Overuse Causes Resistance and Toxicity	26
How To Avoid The Side Effects?	27
Antifertility Drug	28
Tranquillizers	29
Stimulants and depressants	30
Antihistamines	32
Antiseptics and Disinfectants	33
Narcotics	34
Chemotherapy	35
Chapter 3: Chemistry in Personal Care Products and Cosmetics	36
Introduction	36
Lead	38
Ethylene Oxide	38

  Parabens 38
  Phthalates 39
  Triclosan 39
  Chemical Free Cosmetics 39
  Why Chemical Free Cosmetics? 40
  Cost 40
 Hair Dyes: Herbal Vs Chemical 41
  Alternative 42
  How does natural hair dye work? 43
 Anti-aging Creams 44
  What One Should Know About Chemicals in Anti -Aging Creams? 45
  Which Ingredients Should One Look for in an Anti- Aging Skin Care System? 45

Chapter 4: Presence of Chemistry in food and Drinks 47

 Artificial Sweetners 49
 Antioxidants 51
  Uses in technology 52
  Food preservatives 52
  Industrial uses 53
  Levels in food 55
 Natural Alternatives 57
 Beer-A Chemical Drink 58
  Carbohydrate 58
  Hops 59
  Yeast 60
  Water 61
  Carbonation 61
  Nitro beer 62
 Soft Drinks 63
  The Difference between Energy Drinks and Soft Drinks 65

Chapter 5: Chemistry in Toiletries 67

 Introduction 67
  What can you do ? 68
 Natural Vs Chemical Toiletries 70
  Do natural toiletries cost more than chemical based ones? 70
  But what are essential oils? 71
 What are the Benefits of Handmade Soaps and Toiletries? 72

Chapter 6: Chemistry in Agriculture — 74

- Importance of Chemicals in increasing agricultural yield — 74
- What are Pheromones? — 75
- Why is Microbial Fertilizer the Best Option for Agriculture? — 77
- Using Organic Agriculture Products — 79
  - Biopesticides are Likely to Gain Momentum — 80

Chapter 7: Chemistry involvement in The construction industry — 82

- The Chemicals used in Construction — 84
- Constructing Roofs and Walls — 85
  - Constructing Floors — 85
  - Constructing Doors and Windows — 85
  - For Bathroom Fittings — 86
  - Chemicals used as pest control in construction of new house 86

Chapter 8: Various Industrial Chemicals and Their Uses — 88

- Various Types of Industrial Chemicals — 90

Chapter 9: Chemistry in Rocket Propellants/Space Vehicles — 92

- Materials used in Making International Space Station — 94
- Glass used in Spacecrafts — 95

Chapter 10: Chemistry Involved in Clothing — 96

- Examples of clothing materials — 96
- Why Is Natural Organic Clothing Vital For Your Health? — 99
- Different Dyes used to the Fabrics — 101
  - Animal-derived dyes — 102
  - Plant-derived dyes — 102
  - Types — 102

Chapter 11: Chemicals used in Paper Industry — 105

- Various Chemicals used in different steps of paper making — 106
  - Pulping — 106
  - Bleaching — 106
  - Sizing — 107
  - Strengthening — 107
  - Binders — 108
  - Fillers — 108
  - Retention — 109
  - Coating — 109

Chapter 12: Polymers — 110
Organic Polymers — 113
Brand names — 115

Chapter 13: Warfare and Defence — 119
Chemical Warfare — 119
Nuclear Weapons — 122

Chapter 14: Fats and Oils — 124

Chapter 15: New High Performance Materials — 128
Carbon Fibres — 128
Ceramics — 130
Microalloys — 132

Chapter 16: Fuels — 133
Chemical — 134
Biofuel — 135
Fossil Fuels — 135
Nuclear Fuels — 137
Liquid fuels for transportation — 138

Chapter 17: Chemistry at Crime Scene/Forensic Chemistry — 139
Inspection at crime scene — 139
Chemicals used to reveal fingerprints — 140
Taking a Closer Look at the Evidence of a Crime Scene — 142
Interpreting the Results — 145

Chapter 18: Potable Water — 146
Different processes to clean drinking water — 147
Processes — 147
Chlorine dioxide method — 149
Ultraviolet radiation — 149
Ozone purification — 150
Bromination and Iodinization — 150

Chapter 19: Chemical Waste — 151
Chemical Compatibility Guideline — 153
Container compatibility — 153
Requirements for chemical waste disposal — 154
Packaging — 154
Labelling — 154
Storage — 154

Chapter 20: Chemistry inside Human Body — 156

Composition of the Human Body — 156
   Elements — 157
   Molecules — 158
Chemistry in Food Digestion — 160
   Significance of pH — 162

Chapter 21: Chemistry in Cooking — 163

What is Cooking? — 163
   Carbohydrates — 164
   Lipids — 165
   Proteins — 166
   Taste — 167
Chemical Reactions involved in Baking a Cake — 169
   Gluten Formation — 169
   Leavening Agents — 170
   Browning Reactions — 170
   Emulsification and Binding — 171
Onion Chemistry — 172

Chapter 22: Chemistry in Recreation — 173

Chemistry and Sports — 173
Chemistry in Swimming Pools — 177
   Total Alkalinity — 177
   pH Level — 177
   Chlorine Compounds — 178
   Chloramines — 179
   Cyanuric Acid Level — 179
   Amount of Dissolved Solids — 179
   Temperature — 180
Photography — 181
   Black and white negative processing — 182
   Black and white reversal processing — 183
   Colour processing — 184

Chapter 23: Chemistry and Consumer Products — 185

Chemistry in Cells/Batteries — 185
Chemistry and Fireworks — 187
   The chemistry of fireworks colours — 187
Shoe polish — 189
   Types — 189

Wax-based shoe polish	189
Cream-Emulsion shoe polish	189
Liquid shoe polish	190
Chemistry and Sand	191
Chemistry in Safety Matches	193
Working of Safety Matches	193
Mosquito Coils	195
Fountain Pen Inks	196
Crayons	197

Chapter 24: Chemistry and Atmosphere of Earth — 198

Composition	198
Chemical Pollutants in the Environment	200
Anthropogenic (man-made) sources	200
Natural sources	201
"Different Kinds of Pollutants"	203

Glossary — 207

CHAPTER 1

Introduction

Contribution of Chemistry in Improving our daily life

The field of chemistry has always been involved in every aspect of our daily lives. Research in chemistry has led to life changing discoveries, allowed us to progress as a society, has made our lives safer, and even allowed us to live longer. Without chemistry, we would not have nearly all of the products that we wear, eat, and use daily. It not only has improved our daily lives, but has made a significant impact on how our society has evolved and flourished.

Chemical research and development has resulted in improvements in the production of food and water. Without it, there would be much less safe foods to eat and clean water to drink. One of the most important chemical contributions to society is the development of chemical polymers (see chapter 12). Plastics, Nylon, PVC (polyvinyl chloride), silicone, polyester, and polycarbonate, can be found in every part of our lives such as in our homes, schools, buildings, and work place. Developments from chemical research affect where and how we eat and play and it allows us to have many hobbies and interests as well. Paper, wood products, and metals such as steel and aluminum, are essential items developed from chemical research.

Without chemistry, we would not have access to the variety of foods and food ingredients we see in the grocery stores. In the field of medicine, we would not have drugs such as antibiotics, pain relief medications, and medications for illnesses such as asthma, heart disease, and diabetes, etc. The remarkable surgeries being performed today saves millions of lives. Without chemicals used to develop products to allow surgeries to be performed, we would not have access to life saving health care. Such surgical items include: anesthetics, latex gloves, sterilization equipment and solutions, etc. Then all the metabolic processes occurring in our body are nothing but biochemical reactions. The growth of living organisms, the ripening of fruits; the dyes (See chapter10), polymers, paper, etc. that we use in our day to day life, without which our life seems impossible are all results of chemical reactions. The characteristics that we inherit from our parents are, as we all know, due to the DNA. The DNA is nothing but a polymer. Similarly, hair, skin, cotton and silk fibres, cellulose are all polymers. Polymers are nothing but long chains of atoms.

How can we forget the source of life on earth i.e. the Sun, without which there will be no life on earth? It also gets its energy by a chemical reaction called 'Nuclear fusion' where two Hydrogen atoms combine to give a heavier atom i.e. Helium, producing large amounts of energy

Introduction

during the process. Thus, everything surrounding us has got something or the other to do with chemistry.

Do we ever realize how these chemicals combine on their own to give such complex reactions? How the atoms know that they are supposed to combine with this atom or that and in what proportion? Is this not a beautiful aspect of the mystery we call 'Chemistry'? It's a complete mystery as to how these atoms don't even have to think once before they react but if we, human beings, try to make these things artificially, the pain and effort that goes into making it is really enormous. Many a times, the number of attempts that go into making of a product are more than one. A chemist really strives to make products which are useful to us and safe for use for e.g., the medicines; the polymer products like foam mattresses, clothes, furniture, plastic goods, detergents, cosmetics, and other consumer goods; these and many are all gifts of chemistry to mankind.

When it comes to sustaining life on the planet, without chemical research, we would not have essential products as fertilizers, and herbicides and pesticides needed for the agricultural industry. We would also not have the ability to have sewage treatment plants which reduces dangerous illnesses.

Everything we find in our homes is the result of research in chemistry. Chemical processes have resulted in a broad variety of products and materials needed for an ever evolving modern society. For instance, soap, toothpaste, deodorant, shampoo, toothbrushes, shaving supplies, make up, and other personal care products have all been created from chemical research. Kitchen items such as pots, pans, silverware, plates, and cups were created with the help of chemistry. Almost all of the cleaning products found in a home are the result chemical developments. As well, without chemistry, we would not have such items as synthetic fabric, computers, CDs, DVDs, iPods, fuel for vehicles, refrigeration units, radios, televisions, radios, batteries, and so much more.

For centuries, chemical research and discovery has played a fundamental role in improving the quality and extension of life. Research in chemistry is essential to understanding life and the environment. Wherever we are, some part of research in chemistry is touching our lives.

Thus Chemistry, which is so useful to us, is also very harmful to us, if used carelessly. It's a shear mystery that all these gifts of chemistry could be destroyed even before an eye can blink by some other products of chemistry like the dynamite, atom bomb, hydrogen bomb, biochemical weapons, etc.

CHAPTER 2

Chemistry in Medicines and Health Care

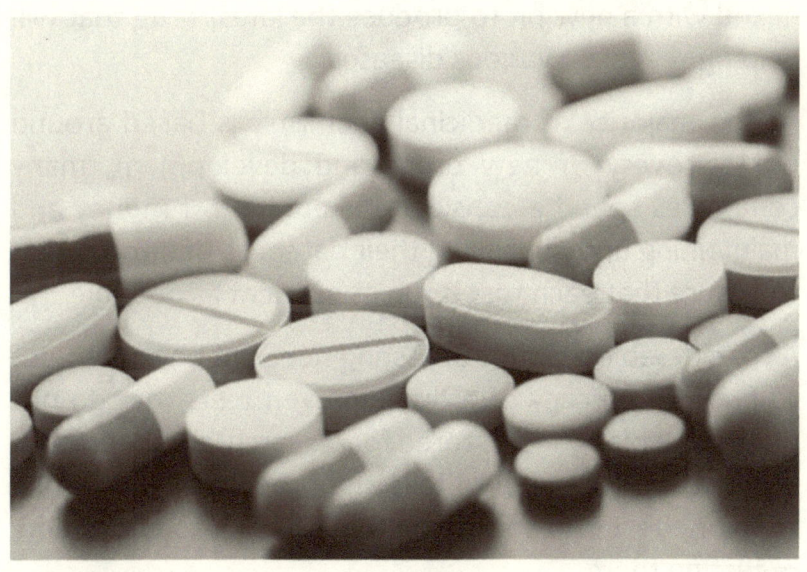

Introduction

Medicinal chemistry is the term used to describe the combination of sciences used to develop the pharmaceutical drugs that stock the shelves of our chemists and hospital departments - and this is its main purpose and achievement. Medicinal chemistry has and will continue to play an important role in today's society as it deals with development, synthesis and design of pharmaceutical drugs. These results are then used to give us a better understanding of diseases as well as giving us ways of preventing and curing them so you can see how important medicinal chemistry is to modern day society and funding for projects that are taking place around the world is vital.

Although medicinal chemistry is about creating new drugs, the properties and quantitative structure activity relationships (QSAR)[1] of existing drugs is important to see if a combination of these biological properties can be mixed with a new hit to produce the latest drug that will help fight against various diseases.

As the majority of medicinal chemistry is based around the discovery of new drugs and development, many companies spend a considerable amount of money and maintaining and improving their database of information to ensure that each test is run as efficient as possible.

A considerable amount of testing needs to be carried out to find the efficacy of a drug so even tests that do not quite go to plan can be stored for research later. Medicinal chemists may be able to take the properties and results from one test into another without having to waste time re-testing them.

With so much data and information required, it is no surprise that medicinal chemistry involves various disciplines including toxicology[2], pharmacology[3], biochemistry[4] and molecular biology[5]. This also shows that those who work on these projects have to be able to communicate well with each other as well as have a good understanding of the various scientific disciplines in order to get the best results.

The drug discovery process begins with the ability to identify novel active compounds that is referred to as hits within the industry. These hits are found by conducting research and can be found through many different sources including fungi, plants and animals so it is important to make sure that the research conducted is thorough.

But as well as drug discovery, medicinal chemistry studies molecular interaction, in other words what happens between molecules in cells in proteins, carbohydrates, lipids and so on. These studies are often performed in order to recognise molecular interactions and then study the effects of these interactions to understand whether or not they can produce interesting and noteworthy results that are essential to the development of new drugs.

Findings from these and other important studies form the basis of much medicinal chemistry literature and it is often this literature that experts in the field turn to when they need information on a particular organic compound, to identify a molecular interaction and much more.

Literature and case studies on the subject of medicinal chemistry are therefore invaluable to those in this industry, and their availability at a moment's notice is also extremely important. Thanks to the internet and to organisations pulling libraries of information together on this discipline, it has never been easier for scientists to benefit from the extensive work of their fellows and to use past research to help with future studies.

As well as the results of the medicinal chemistry studies being published in journals and research papers, methods, strategies and targets are also discussed in such documents, and these pieces of information can help people to understand the direction of the discipline and what they need to be achieving. This knowledge-sharing practice helps to avoid repetition and move the process of drug discovery forwards at a steady pace.

Antacids

An **antacid** is a substance which neutralizes stomach acidity and is used to relieve heartburn, indigestion or an upset stomach.

When excessive amounts of acids are produced in the stomach, the natural mucous barrier that protects the lining of the stomach can damage the esophagus in people with acid reflux[6]. Antacids contain alkaline ions that chemically neutralize stomach gastric acid, reducing damage and relieving pain.

Digene is the number 1 doctor prescribed antacid in India and is trusted by Indian consumers for over the past 70 years to provide fast, symptomatic relief from heartburn and gas.

Examples of antacids include: Alka-Seltzer, Milk of Magnesia, Alternagel, Amphojel, Gaviscon, Gelusil, Maalox, Mylanta, Rolaids, Pepto-Bismol, Tums etc.

Analgesics

An **analgesic** or **painkiller** is any member of the group of drugs used to achieve analgesia, relief from pain. They are also known as painkillers or pain relievers. Technically, the term **analgesic** refers to a medication that provides relief from pain without putting you to sleep or making you lose consciousness.

Analgesic drugs act in various ways on the peripheral and central nervous systems. They are distinct from anesthetics, which temporarily affect, and in some instances completely eliminate, sensation. Analgesics include paracetamol (known in North America as acetaminophen or simply APAP), the nonsteroidal anti-inflammatory drugs (NSAIDs) such as the salicylates, and opioid drugs such as morphine and oxycodone.

Antibiotics

Antibiotics are among the most commonly prescribed medications in modern medicine. The word antibiotic is composed of two words, anti means 'against' and bios means 'life'. Antibiotics are also called antibacterial, and they are drugs specifically used to treat infections caused by bacteria; it is essential to understand that antibiotics only treat bacterial infections. Antibiotics are ineffective against viral infections such as the common cold and fungal infections such as ringworm. Bacteria are very tiny organisms that can at times cause illness to both humans and animals. Antibiotics treat the diseases by killing or destroying bacteria. The first antibiotic in the history of medicine was penicillin, which was discovered out of the blue from a mold culture[7]. In our modern world today, over hundreds of different antibiotics are available to cure from minor to life-threatening infections, such as tuberculosis, salmonella, syphilis and some forms of meningitis. Penicillin-related antibiotics are ampicillin, amoxicillin and benzyl penicillin, which are extensively used nowadays to treat a variety of infections; these antibiotics or antibacterial have been around for a long time.

When To Take Them?

Antibiotics do not work for every illness, as it meant to cure the infections caused by bacteria, so one must consider this fact while taking any antibiotic. Doctor's prescription is very important for taking antibiotics because a doctor can tell well that if a patient has viral infection or bacterial infection. Antibiotics should not be taken for viral illnesses, such as colds or the flu, because by taking antibiotics in viral infections one makes the illness worse as it enables the bacteria to resist the antibiotics. Take the antibiotics on Doctor's professional prescription so that one get benefit from it, the dosage of the antibiotics should be taken properly and regularly because once patient miss any dose then it can result in the resistance from bacteria, which will make the whole treatment ineffective.

Some of the symptoms of onset of any bacterial infection are red throat and tonsils, difficulty in swallowing, fever ranges above 101 degrees, swollen and tender lymph nodes in the neck, headache, shivers and shaking with cold sweats, often nausea, vomiting and abdominal pain in children.

Overuse Causes Resistance and Toxicity

The overuse of antibiotics causes "Antibiotic resistance" and "bacterial resistance". Frequently, antibiotics destroy bacteria or simply stop them from growing; however, some of the bacteria have become resistant to some types of antibiotics. This means that the antibiotics no longer work against such bacteria. Bacteria become resistant to antibiotics more quickly when antibiotics are used too often or are not used properly, for example, if a person does not take the full course of a specific antibiotic prescribed by any doctor, then the bacteria in his body will develop a resistance against that specific antibiotic, such bacteria can, then be treated with some other antibiotic, but some types of bacteria are resistant to all the antibiotics and unable to get treated. Moreover, the overuse of antibiotics can also cause toxicity in the body at different cellular, tissue or organ levels. For example, Quinolones are a class of antibiotics that are very toxic for tendons[8], cartilages[9], nervous system[10] and various

Chemistry in Medicines and Health Care

other organs. The overuse results in the accumulation of antibiotics in lysosomes which may cause metabolic alterations that can lead to cell toxicity.

How To Avoid The Side Effects?

To avoid the side effects of antibiotic's overuse, one must avoid the overuse of antibiotics at first place. To control the antibiotic overuse one must not depend on the doctor, regulate the antibiotic use by yourself. Following is some of the suggestions to avoid the resistance of antibiotics:

- ❖ Don't always follow the urge of asking your doctor for antibiotics whenever you have flu, cough or cold. Bacterial infections generally go away on their own within two weeks.
- ❖ While taking any antibiotic drug from doctor, always ask him if it is necessary in the present condition or not and also that you have really a bacterial infection or viral one.
- ❖ Always follow the doctor's direction of the use of antibiotics.
- ❖ Always make sure to complete the full course of prescribed antibiotics, even if you feel recovered in the mid of the course because if you left taking it in the mid of the course then there is a possibility that the bacteria may get a chance to grow again and develop a resistance, and then you will be supposed to take stronger antibiotics to recover that second attack.
- ❖ Take proper and healthy diet with antibiotics so that your immune system grows stronger and able to resist the infections

Antifertility Drug

The **combined oral contraceptive pill (COCP)**, often referred to as the **birth control pill** or colloquially as **"the pill"**, is a type of birth control that is designed to be taken orally by women. It includes a combination of an estrogen (usually ethinylestradiol) and a progestogen (specifically a progestin). When taken correctly, it alters the menstrual cycle to eliminate ovulation and prevent pregnancy.

Tranquillizers

A tranquilizer refers to a drug which is designed for the treatment of anxiety, fear, tension, agitation, and disturbances of the mind, specifically to reduce states of anxiety and tension.

Tranquilizers fall into two main classes, major and minor. Major tranquilizers, which are also known as antipsychotic agents, or neuroleptics, are so called because they are used to treat major states of mental disturbance in schizophrenics and other psychotic patients. By contrast, minor tranquilizers, which are also known as antianxiety agents, or anxiolytics, are used to treat milder states of anxiety and tension in healthy individuals or people with less serious mental disorders.

The principal minor tranquilizers are the benzodiazepines, among which are diazepam (Valium), chlordiazepoxide (Librium), and alprazolam (Xanax). These drugs have a calming effect and eliminate both the physical and psychological effects of anxiety or fear. Besides the treatment of anxiety disorders, they are widely used to relieve the strain and worry arising from stressful circumstances in daily life. Benzodiazepines work by enhancing the action of the neurotransmitter gamma-aminobutyric acid (GABA), which inhibits anxiety by reducing certain nerve-impulse transmissions within the brain.

Stimulants and depressants

A depressant, or central depressant, is a drug that lowers neurotransmission levels, which is to depress or reduce *arousal* (Arousal is the physiological and psychological state of being awaken or of sense organs stimulated to a point of perception.) or stimulation, in various areas of the brain. Depressants are also occasionally referred to as *"downers"* as they lower the level of arousal when taken. They slow the messages going to and from your brain. In small quantities depressants can cause a person to feel relaxed and less inhibited. In large amounts they may cause vomiting, unconsciousness and death. Examples of depressants include: alcohol, opiates (such as heroin and morphine), cannabis, sedatives (such as Valium) etc.

Stimulants or "uppers" increase mental and/or physical function, hence the opposite drug class of depressants is stimulants, not antidepressants. Stimulants are useful in treating many medical conditions, including ADHD[11], narcolepsy, asthma, obesity, and depression. There are several types of stimulants, including caffeine, nicotine, cocaine, amphetamines, methamphetamines, dextroamphetamine (Dexedrine, Dextrostat, ProCentra), methylphenidate (Concerta, Daytrana, Methylin, Ritalin), and the combination of amphetamine and dextroamphetamine (Adderall). Caffeine is a stimulant compound belonging to the xanthine class of chemicals naturally found in coffee, tea, and (to a lesser degree) cocoa or chocolate. It is included in many soft drinks, as well as a larger amount in energy drinks. Caffeine is the world's most widely used psychoactive drug and by far the most common stimulant. Caffeine is also included in some medications, usually for the purpose of enhancing the effect of the primary ingredient, or reducing one of its side-effects (especially drowsiness).

Roasted coffee beans –a common source of Caffeine.

Antihistamines

Antihistamines are drugs which treat allergic rhinitis[12] and other allergies. Typically people take antihistamines for getting relief from nasal congestion, sneezing, or hives[13] caused by pollen, dust mites, or animal allergy with few side effects.

Antihistamines are usually for short-term treatment. Chronic allergies increase the risk of health problems which antihistamines might not treat, including asthma, sinusitis, and lower respiratory tract infection.

Antihistamines are a class of agents that block histamine release from histamine-1 receptors and are mostly used to treat allergies or cold and flu symptoms. Histamine-1 receptors are located in the airways, blood vessels and gastrointestinal tract (stomach and esophagus). Stimulation of these receptors can lead to conditions such as a skin rash or inflammation, a narrowing of the airways (bronchoconstriction), hay fever, or motion sickness. Histamine-1 receptors are also found in the brain and spinal cord, and stimulation of these receptors makes you more awake and alert. Sedating antihistamines oppose the effects of histamine on H1 receptors in your brain, which is why they cause sedation and drowsiness.

Antiseptics and Disinfectants

Antiseptics are antimicrobial substances that are applied to living tissue/skin to reduce the possibility of infection, sepsis[14], or putrefaction[15]. Antiseptics are generally distinguished from *antibiotics* by the latter›s ability to safely destroy bacteria within the body, and from *disinfectants*, which destroy microorganisms found on non-living objects.

Some antiseptics are true *germicides*, capable of destroying microbes (bacteriocidal), while others are bacteriostatic and only prevent or inhibit their growth.

Antibacterials include antiseptics that have the proven ability to act against bacteria. Microbicides which destroy virus particles are called viricides or antivirals. Antifungals, also known as an antimycotics, are pharmaceutical fungicides used to treat and prevent mycosis (fungal infection).

Narcotics

The term **narcotic** originally referred medically to any psychoactive compound with sleep-inducing properties. In the United States, it has since become associated with opiates and opioids, commonly morphine and heroin, as well as derivatives of many of the compounds found within raw opium latex. The primary three are morphine, codeine, and thebaine (while thebaine itself is only very mildly psychoactive, it is a crucial precursor in the vast majority of semi-synthetic opioids, such as oxycodone).

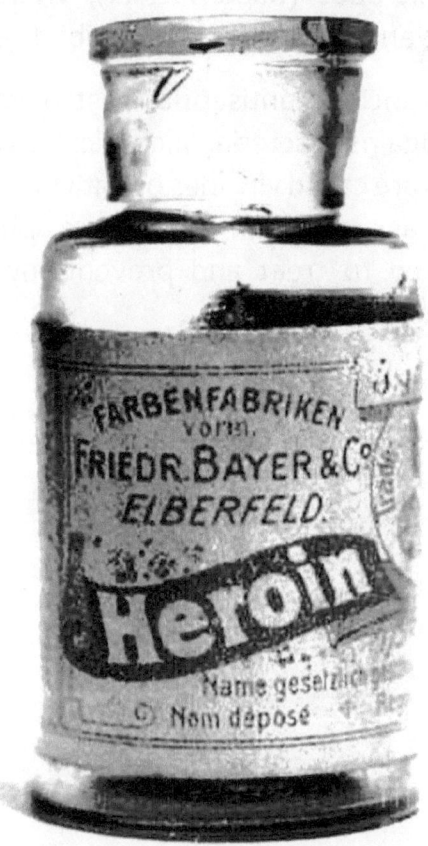

Heroin, a powerful opioid and narcotic

Chemotherapy

Chemotherapy, in its most general sense, refers to treatment of disease by chemicals that kill cells, specifically those of micro-organisms or cancer. Chemotherapy works by destroying cancer cells; unfortunately, it cannot tell the difference between a cancer cell and some healthy cells. Chemotherapy may be given in many ways. It can be administered through a vein, injected into a body cavity, or delivered orally in the form of a pill, depending on which drug is used. Chemotherapy is sometimes used along with other cancer treatments, such as radiation therapy, surgery, or biological therapy (the use of substances to boost the body's immune system while fighting cancer). Chemotherapy was formed from mustard gases[16], which was in use as chemical- arms during the 1st World-War. Chemotherapy for the treatment of cancer began in the 1940s with the use of nitrogen mustard.

The course of therapy will depend on the cancer type, the chemotherapy drugs used, the treatment goal and how your body responds. Adjuvant chemotherapy[17] and radiotherapy[18] are often given following surgery for many types of cancer, including colon cancer, lung cancer, pancreatic cancer, breast cancer, prostate cancer, and some gynaecological cancers. Over time, cancer cells become more resistant to chemotherapy treatments.

CHAPTER 3

Chemistry in Personal Care Products and Cosmetics

Introduction

Did you know that we come into contact with 100,000 chemicals every day? And less than 5% of these chemicals have been thoroughly tested for their long term effects on humans? Those are pretty scary numbers for those of us striving to live greener lives whether by purchasing organic clothing or recycling. Head over to your medicine cabinet (or wherever you keep your makeup, your skincare goods, your shampoos), and read the ingredient labels. One facet of our lives we forget to make greener is what we put on our skin every single day.

Cosmetics are not regulated by governmental agencies. That shampoo you put in your hair this morning and that moisturizer you put on your face? Their safety is determined by the cosmetic manufacturers. Does that thought scare you?

Thankfully, you will find some things you can do. First, don't buy goods that contain phthalates, formaldehyde, phenols, sodium laureth sulfate, coal tar, toxic dyes, and synthetic fragrances. Next, do not believe that a product is "natural", because it is labeled as such. There are numerous goods out there labeled as "natural" and "organic" that still contain paraben and formaldehyde preservations.

An additional factor to bear in mind is you will NOT look like a celebrity. Don't fall into the advertising trap and acquire a product just for the reason that your favorite celebrity wears it. Photoshop is their very best friend. If you should see a great deal of the celebrities out there without makeup, they look more like us than you'd believe.

Many cosmetics contain dangerous chemical ingredients that can cause damage much deeper than just skin level. The cosmetics industry adds synthetic chemicals to anything from shaving cream to body lotion to lipstick. Many of these ingredients are the same ones that can be found in products used to stabilize pesticides and clean industrial equipment. You may think that it would be illegal to include such dangerous ingredients in cosmetics that are sold to the public, but major loopholes in federal laws allows the practice to continue, despite these ingredients being linked to birth defects, infertility, and even cancer.

Many individuals do not thoroughly understand the dangers in chemical based cosmetics, and never give a second thought to what they may be applying to their skin. Remember, the skin is the largest bodily organ. Whatever is applied to it is absorbed directly into the bloodstream.

The continuous use of chemical based cosmetics has been linked to numerous adverse health effects, yet most people continue to use them without realizing just how dangerous they really are.

Below, we will discuss a few of the most-used chemicals in cosmetics, and how they affect your health:-

Lead

Lead has been found in over 650 various cosmetics, including lipstick, nail polish, bleaching toothpaste, foundation and sunscreen. Lead is a known neurotoxin that has been linked to behavioral problems, speech delays, and learning disabilities in children. It has been shown to delay puberty[19] in girls and reduce the fertility of both men and women.

Ethylene Oxide

Ethylene oxide is a chemical substance that is traditionally used to sterilize surgical instruments. Because it has been shown to buffer the effects of sudsing[20] agents, trace amounts of it can be found in body washes, shampoos and facial cleansers. According to the National Toxicology Program, it is also classified as a carcinogen.

Parabens

Parabens are one of the most commonly found chemicals in chemical-based cosmetics. They are used in deodorants, creams, lotions, and ointments as a type of preservative. They are very easily absorbed by the skin, making them one of the most dangerous chemical ingredients known. In fact, they have even been identified in samples of breast tumors.

Phthalates

Phthalates are chemicals found in synthetic fragrances (such as scented body sprays) and nail polish which have been shown to disrupt the endocrine system. This can cause girls to experience puberty early, which increases their risk of developing breast cancer later in life.

Triclosan

Triclosan is a popular chemical used extensively in toothpastes, deodorants, and liquid antibacterial soaps to prevent the growth of mold. This chemical is technically classified as a pesticide. Studies have shown that using products that contain this chemical can disrupt thyroid function and hinder normal breast development. Many hand sanitizers also contain triclosan, and have been proven to contribute to bacterial resistance.

Although most cosmetics contain some form of chemical, there is something you can do to avoid using them. Make sure the cosmetics you use contain only natural ingredients, preferably harvested from organically grown and GMO-free sources. What you put IN your body is important for optimal health, but what you put ON your body is just as important.

Chemical Free Cosmetics

When you choose chemical free cosmetics, you do something good not only for yourself, but for the planet as well. Mineral makeup, vegetable-based cosmetic and beauty products and other natural cosmetic products are an excellent way to avoid poisoning yourself and the planet. In addition, such mineral-based chemical free cosmetics are also cruelty-free-no animal testing is done!

Why Chemical Free Cosmetics?

At one time, it may not have mattered so much. Today however, with so many toxic substances in the air, water and food supply, people are increasingly finding their resistance to such poisons strained almost to the breaking point. Using cosmetic products derived from natural, minimally processed mineral and plant sources means one less assault on your bodily systems.

The key word here is "processing." The materials that are used in many commercially mass-produced cosmetic and beauty products often start out as harmless, natural substances that are not that different from those used in organic mineral makeup. The difference, just as with food, is processing which removes many desirable substances and adds many others that we would be better off without.

Cost

As you might imagine, chemical free cosmetics are more expensive than commercially produced cosmetic products. However, in terms of your health and that of the planet, the purchase and use of chemical free cosmetics is an incredible investment - one that will pay off well for future generations.

Hair Dyes: Herbal Vs Chemical

Have you ever thought about the risk you are taking when you sit in hairdresser's chair while your hair is being coated with a chemical hair dye? At the moment it is estimated that more than 1/3 of women over age 18 and about 10 percent of men over age 40 use some type of chemical hair dye.

Conventional permanent hair dyes now make up about 80 percent of the whole hair colours market. They consist of colourless dye "intermediates" (aromatic amines) and dye "couplers."

When the active ingredient - either hydrogen peroxide or ammonia is added, the intermediates and couplers react with one another to form pigment molecules. Darker shades require the higher concentrations of intermediates.

Generally over 5,000 different chemicals are used in hair dye products. Among them: aromatic amines - are suspected of causing cancer, frequently trigger allergies (toluylene diamine, phenylendiamine: "developers"), sodium picramate: "color reinforcer"); ammonia (makes the outer layer swell up, stresses the hair); hydrogen peroxide (lightener); resorcin (connects itself to the developer molecules, thereby determining the colour; can provoke allergic reactions); azo dyes (CI 11680 to CI 40215); polyethylene glycols (PEG; can make the skin more permeable to foreign substances); diethyl phthalate (DEP); synthetic fragrances which frequently trigger allergies.

Some of the chemical are proved to be carcinogenic. Chemical hair dye has been linked to a range of cancers, including tumours of the breast, ovaries, bladder, brain and leukaemia. It is not only cancer that people are worried about-the allergy to the contents of the hair dyes has becoming more and more widespread and may lead to fatal results.

So what happens to your hair when it is being treated with a permanent hair dye?

1. First of all your hair structure is broken open
2. Then all natural pigments are removed
3. Finally they are replaced by synthetic pigments

This results in a complete alteration of your initial hair colour, the damage to the upper hair cuticle layer. As the consequence your hair becomes brittle and lifeless. Moreover, the chemicals called aromatic amines penetrate the skin and stay on the hair for weeks, months and even years. As the time passes, they could react with tobacco, smoke and pollution fumes to form extremely poisonous chemicals called N-nitrosamines.

Alternative

Countless women around the world have achieved breathtaking hair colour results with natural herbal hair colours. It is a true Nature's gift that is pure botanical and represents the gold standard of the mix-it-yourself hair dyes. Ranging from a luminous golden blonde to a shimmery black, organic herbal hair powders or creams feature a mixture of 100% natural colorants, enriched with conditioning wheat protein and jojoba oil. The colorants normally include certified organically grown henna, walnut shells, buckthorn, cassia, indigo, hibiscus, rhatany, coffee, rhubarb, curcuma, and beetroot. The conditioners are powdered wheat protein and jojoba. Additionally, the products contain essential oils to improve the fragrance. These hair dyes are primarily intended for home use. The countless amounts of users tend to combine several colours to achieve their own unique hair shade. If it is a powder it can be mixed with either water, black tea, coffee or even red wine. The creams are already premixed and ready to be applied directly out of the tube.

How does natural hair dye work?

1. First of all the colours do not interfere chemically with natural hair structure
2. The plant-sourced pigments attach themselves to the hair's surface and cover it like a protecting glaze
3. The result is brilliance of colour, sheen and volume, strengthening of hair structure and smoothing of hair surface.

All in all your new hair colour results from the combination of your initial colour with the applied hair colour layer. Each time it can be unique and special. In general, natural herbal hair dyes are very suitable to colour gray hair. However, purely natural colorants will never make the gray "disappear" in the manner of a chemical process. Due to its unique characteristics, gray hair will always produce a lighter result than non gray hair when coloured with a single application of a hair colour. The more gray you have, the lighter the overall colour result will be. The different colour gradations in your hair, including the gray, can produce fascinating multi-coloured results.

It is obvious that there is no need to risk your own health. And it does not mean you have to be a gray mouse - discover the great variety of herbal hair dyes and find your own unique shade.

Anti-aging Creams

As medical researchers have worked to unlock the mysteries of the aging process one area of research that has yielded a high rate of success was in understanding the aging process of the skin. This is because unlike other organs of the body, the skin is easily accessible for study and also for the same reason it is just as easily accessed for treatment.

It was discovered through intensive research that there is a myriad of biological functions that transpire below the skins surface in the substrate that directly effect the way that the skin looks on the surface. One important discovery that was made is that the skin produces a myriad of substances such as peptides, hyaluronic acid, collagen and elastin that it uses to maintain its overall health and appearance.

However; as we age, these substances are produced in lesser and lesser amounts which consequently has a direct negative effect on the way the the skin on the elderly looks and feels.

The good news is that new anti- aging skin care products that are now available contain chemically engineered substances that are able to penetrate below the skins outer protective surface layer to the substrate where they are then able to function as a replacement therapy for these necessary substances that are deficient in the elderly.

The actual cause of skin wrinkles lies below the skins surface and this is where any treatment that is to have any effect on skin wrinkles must take place. What these new and highly effective anti- aging skin care products do is to stimulate the actual living skin cells in the substrate to produce more of the substances that are needed for healthy and younger looking skin.

There are so many anti- aging products and anti -aging review sites that it can be hard to know which product is the best.

Generally speaking, the best skin creams are the ones that use all natural ingredients. Unfortunately, many of the most expensive creams in the market use chemicals that have been shown to be harmful to a person's health.

What One Should Know About Chemicals in Anti-Aging Creams?

Here are a few facts that most anti -aging skin care review sites do not tell people:

- The FDA does not regulate the cosmetic industry. In fact, there is no government body that tells the industry which ingredients it can and cannot use in their skin creams or lotions.
- Many anti- aging products contain chemicals that are known carcinogens. Some of these harmful chemicals include Dioxane, Mineral Oils, Alchohols, Parabens and Fragrances.
- The chemicals in any skin cream will seep through the skin and enter the body.
- Many commercial creams contain allergens that have a negative impact on those with sensitive skin.

Which Ingredients Should One Look for in an Anti-Aging Skin Care System?

Many anti-aging skin care review sites focus on brand

names, not ingredients. However, the ingredients used in the skin cream are the most important factor. There are certain natural ingredients that have a tremendously positive effect on a person's skin. These ingredients should be near the top of the ingredient list in any good anti-aging product:

- **Nano-Lipobelle H-EQ10:** This substance is a special nano-form of the antioxidant Coenzyme Q10. Coenzyme Q10 is produced naturally by a person's body. Unfortunately, the older a person gets, the less of this substance is produced. Nano-Lipobelle H-EQ10 is an improved form of COQ10 which is a powerful antioxidant that will not only reduce wrinkles but also prevent them from occurring in the first place.

- **Phytessence Wakame:** Phytessence Wakame is a kelp plant that is native to Japan. This natural substance gives the skin a firmer look and feel. It contains numerous minerals that will keep the skin looking healthy and young.

- **Xtend-TK:** This substance has been clinically proven to improve elasticity in the skin. It is not instantaneously effective but it does work very well. Xtend-TK is a cutting edge substance that enables the skin to repair.

Choosing the right anti-aging skin care system is important. No one finds wrinkles beautiful. Although wrinkles are a natural part of the aging process, a person who chooses the right skin care system will find that many wrinkles can be reduced or even eliminate.

CHAPTER 4

Presence of Chemistry in food and Drinks

Chemicals are essential building blocks for everything in the world. All living matter, including people, animals and plants, consists of chemicals. All food is made up of chemical substances. Chemicals in food are largely harmless and often desirable–for example, nutrients such as carbohydrates, protein, fat and fibre are composed of chemical compounds.

Chemicals can, however, have a variety of toxicological properties, some of which might cause effects in humans and animals. Usually, these are not harmful unless we are exposed to them for a long time and at high levels. Scientists help to safeguard against these harmful effects by establishing safe levels. This scientific advice informs decision-makers who regulate the use of chemicals in food or seek to limit their presence in the food chain.

Chemical substances can play an important role in food production and preservation. Food additives can, for example, prolong the shelf life of foods; others, such as colours, can make food more attractive. Flavourings are used to make food tastier. Food supplements are used as sources of nutrition.

Presence of Chemistry in food and Drinks

Artificial Sweetners

Artificial sweeteners are one way to reduce added sugar in foods. Compounds *like saccharin, aspartame, and sucralose* are hundreds of times sweeter than sugar, so less is needed to generate the same sweet taste. The tiny amount of added artificial sweetener essentially does not increase the calorie count of food. Artificial sweeteners, however, need some chemical help to be palatable. Companies mix them with additives to mask bitter flavors, and they also blend artificial sweeteners to get the best attributes of each one.

Companies searching for new sweeteners include a South Korean company making tagatose, a structural isomer of fructose with roughly the equivalent sweetness of sugar, but with one-third of the calories. The structure of tagatose prevents it from being fully metabolized. Tagatose can be prepared through a two-step synthesis. First lactose is hydrolyzed into glucose and galactose. Then an enzyme converts the galactose to tagatose. The sugar derivative appears to function well in ice cream and baked goods.

Artificial sweetners also do not increase blood glucose levels, which can help people with diabetes manage their condition. However, some consumers try to avoid products containing artificial sweeteners. These additives have faced public scrutiny due to concerns over safety and carcinogenicity[21].

To boost consumer comfort without adding calories, companies are looking for naturally occurring non-calorie sweeteners. One such example available on grocery shelves is *stevia*, a product containing plant-derived glycosides. Scientists have started a pilot plant for producing stevia glycosides by fermentation in engineered microbes. This production method means that the companies can synthesize only the sweetest sugars present in stevia, which are in low concentration in the natural extract, and possibly reduce the product's bitter aftertaste.

Bio Vittoria has teamed up with Tate & Lyle to create a different artificial sweetener, derived from a fruit, not a plant. A representative from a market research firm said this product may gain more attention because it's easier for consumers to connect sweeteners to fruit.

Despite their negligible effect on calories, artificial sweeteners are not necessarily a solution to obesity, as studies indicate that humans and laboratory animals tend to overeat products containing artificial sweeteners. Sugary products, on the other hand, trigger a cascade of hormones that signal the brain that the gut is full.

Antioxidants

Antioxidants are compounds that inhibit oxidation. Oxidation is a chemical reaction that can produce free radicals, thereby leading to chain reactions that may damage the cells of organisms. Antioxidants such as thiols or ascorbic acid (vitamin C) terminate these chain reactions. To balance the oxidative stress, plants and animals maintain complex systems of overlapping antioxidants, such as glutathione and enzymes (e.g., catalase and superoxide dismutase), produced internally, or the dietary antioxidants vitamin C and vitamin E.

The best way to combat free radicals is to have a diet of fresh fruits and vegetables, red wine, and green tea. These functional foods are rich in phytochemicals with antioxidant properties. Health supplements enriched with antioxidants are also now widely available.

The term "antioxidant" is mostly used for two entirely different groups of substances: industrial chemicals that are added to products to prevent oxidation, and naturally occurring compounds that are present in foods and tissue. The former, industrial antioxidants, have diverse uses: acting as preservatives in food and cosmetics, and being oxidation-inhibitors in fuels.

Structure of antioxidant glutathione

Uses in technology

Food preservatives

Antioxidants are used as food additives to help guard against food deterioration. Exposure to oxygen and sunlight are the two main factors in the oxidation of food, so food is preserved by keeping in the dark and sealing it in containers or even coating it in wax, as with cucumbers. However, as oxygen is also important for plant respiration, storing plant materials in anaerobic conditions produces unpleasant flavors and unappealing colors. Consequently, packaging of fresh fruits and vegetables contains an ~8% oxygen atmosphere.

Antioxidants are an especially important class of preservatives as, unlike bacterial or fungal spoilage, oxidation reactions still occur relatively rapidly in frozen or refrigerated food.

These preservatives include natural antioxidants such as ascorbic acid (AA, E300) and tocopherols (E306), as well as synthetic antioxidants such as propyl gallate (PG, E310), tertiary butylhydroquinone (TBHQ), butylated hydroxyanisole (BHA, E320) and butylated hydroxytoluene (BHT, E321).

The most common molecules attacked by oxidation are unsaturated fats; oxidation causes them to turn rancid. Since oxidized lipids are often discolored and usually have unpleasant tastes such as metallic or sulfurous flavors, it is important to avoid oxidation in fat-rich foods.

Thus, these foods are rarely preserved by drying; instead, they are preserved by smoking[22], salting or fermenting. Even less fatty foods such as fruits are sprayed with

Presence of Chemistry in food and Drinks

sulfurous antioxidants prior to air drying. Oxidation is often catalyzed by metals, which is why fats such as butter should never be wrapped in aluminium foil or kept in metal containers.

Some fatty foods such as olive oil are partially protected from oxidation by their natural content of antioxidants, but remain sensitive to photooxidation. Antioxidant preservatives are also added to fat based cosmetics such as lipstick and moisturizers to prevent rancidity[23].

Industrial uses

Substituted phenols and derivatives of phenylenediamine are common antioxidants used to inhibit gum formation in gasoline (petrol).

Antioxidants are frequently added to industrial products. A common use is as stabilizers in fuels and lubricants to prevent oxidation, and in gasolines to prevent the polymerization that leads to the formation of engine-fouling residues.

Antioxidant polymer stabilizers are widely used to prevent the degradation of polymers such as rubbers, plastics and adhesives that causes a loss of strength and flexibility in these materials.

Other polymers susceptible to oxidation include polypropylene and polyethylene. The former is more sensitive owing to the presence of secondary carbon atoms present in every repeat unit. Attack occurs at this point because the free radical formed is more stable than one formed on a primary carbon atom.

Some of the fuel additives and their applications are discussed below:-

Fuel additive	Components	Applications
AO-22	N,N'-di-2-butyl-1,4-phenylenediamine	Turbine oils, transformer oils, hydraulic fluids, waxes, and greases
AO-24	N,N'-di-2-butyl-1,4-phenylenediamine	Low-temperature oils
AO-29	2,6-di-tert-butyl-4-methylphenol	Turbine oils, transformer oils, hydraulic fluids, waxes, greases, and gasolines
AO-30	2,4-dimethyl-6-tert-butylphenol	Jet fuels and gasolines, including aviation gasolines
AO-31	2,4-dimethyl-6-tert-butylphenol	Jet fuels and gasolines, including aviation gasolines

AO-32	2,4-dimethyl-6-tert-butylphenol and 2,6-di-tert-butyl-4-methylphenol	Jet fuels and gasolines, including aviation gasolines
AO-37	2,6-di-tert-butylphenol	Jet fuels and gasolines, widely approved for aviation fuels

Levels in food

Fruits and vegetables are good sources of antioxidant vitamins C and E

Antioxidant vitamins are found in vegetables, fruits, eggs, legumes and nuts. Vitamins A, C, and E can be destroyed by long-term storage or prolonged cooking. The effects of cooking and food processing are complex, as these processes can also increase the bioavailability of antioxidants, such as some carotenoids in vegetables. Processed food contains fewer antioxidant vitamins than fresh and uncooked foods, as preparation exposes food to heat and oxygen.

Antioxidant vitamins	Foods containing high levels of antioxidant vitamins
Vitamin C (ascorbic acid)	Fresh or frozen fruits and vegetables
Vitamin E (tocopherols, tocotrienols)	Vegetable oils, nuts, and seeds
Carotenoids (carotenes as provitamin A)	Fruit, vegetables and eggs

Other antioxidants are not obtained from the diet, but instead are made in the body. For example, ubiquinol (coenzyme Q) is poorly absorbed from the gut and is made through the mevalonate pathway. Another example is glutathione, which is made from amino acids. As any glutathione in the gut is broken down to free cysteine, glycine and glutamic acid before being absorbed, even large oral intake has little effect on the concentration of glutathione in the body. Although large amounts of sulfur-containing amino acids such as acetylcysteine can increase glutathione, no evidence exists that eating high levels of these glutathione precursors is beneficial for healthy adults.

Presence of Chemistry in food and Drinks

Natural Alternatives

Consumer concern over the safety of additives has companies looking for natural colors, flavors, and preservatives. Vanillin is the component in vanilla beans that is responsible for most of their flavor. There is 400-500 times more synthetic vanillin produced than natural vanilla extract. That means that the majority of vanilla flavored or scented products contain vanillin. However, demand for natural sources of vanilla have led chemical companies to search for ways to make vanillin through fermentation. The Belgian company Solvay, which produces most of the synthetic vanillin available today, is developing a process starting from a byproduct of rice bran oil refining. Swiss biotech firm Evolva begins their process with sugar. Both companies are working to scale up production for each of their respective products.

Naturex, a company specializing in natural additives, has been making a natural antioxidant from *rosemary* since 1992. Originally, the extract was used to flavor meat, and then some companies realized it also stabilized the meat. Rosmarinic acid and carnosic acid, the natural antioxidants in rosemary, do not carry the flavor of rosemary. (Naturex is now looking to combine the rosemary compounds with organic acids produced by fermentation to create a product that prevents fat oxidation, as well as microbial growth. Chemistry can help identify new antimicrobial preservatives as well. Chemists first identify the specific microbes that grow on a particular food, then use high throughput screening to identify compounds that kill the microbes or slow their growth.

Beer-A Chemical Drink

The chemical compounds in beer give it a distinctive taste, smell and appearance. The majority of compounds in beer come from the metabolic activities of plants and yeast and so are covered by the fields of biochemistry and organic chemistry.

Do you know what you are putting into your body when you drink beer? Here is a quick list of the main ingredients used to make beer in the process of brewing :-

Carbohydrate

The carbohydrate source is an essential part of the beer because unicellular yeast organisms convert carbohydrates into energy to live. Yeast metabolize the carbohydrate source to form a number of compounds including ethanol. The process of brewing beer starts with malting and mashing, which breaks down the long carbohydrates in the barley grain into more simple sugars. This is important because yeast can only metabolize very short chains of sugars. Long-carbohydrates are polymers, large branching linkages of the same molecule over and over. In the case of barley, we mostly see polymers called amylopectin and amylose which are made of repeating linkages of glucose. On very large time-scales (thermodynamically) these polymers would break down on their own, and there would be no need for the malting process.The process is normally sped up by heating up the barley grain. This heating process activates enzymes called amylases. The shape of these enzymes, their active site, gives them the unique and powerful ability to speed these degradation reactions to over 100,000 times faster. The reaction that takes place at the active site is called

Presence of Chemistry in food and Drinks

a hydrolysis reaction, which is a cleavage of the linkages between the sugars. Repeated hydrolysis breaks the long amylopectin polymers into simpler sugars that can be digested by the yeast. The exact type of malt used to make a beer often depends upon what is accessible from the local area in which the beer is being brewed. Some commonly used malts today include Barley, Wheat, Rice, Oats and Rye Amylopectin consists of many glucose molecules linked together either by 1,6 or 1,4 linkages.

Hops

Hops are the flowers of the hops plant *Humulus lupulus*. These flowers contain over 250 essential oils, which contribute to the aroma and non-bitter flavors of beer. However, the distinct bitterness comes from a family of compounds called alpha-acids (also called humulones) and beta-acids (also called lupulones). Generally, brewers believe that α-acids give the beer a pleasant bitterness whereas β-acids are considered less pleasant.

Yeast

D-Glucose → Glycolyse

Pyruvate → Décarboxylase du pyruvate (−CO_2, +H^+)

Acétaldéhyde → Alcool déshydrogénase (NADH+H^+ / NAD^+)

Éthanol

"Chemical structures showing ethanol fermentation"

In beer, the metabolic waste of yeast is a significant factor. In aerobic conditions, the yeast will use the simple sugars from the malting process in glycolysis, and send the major organic product of glycolysis (pyruvate) into carbon dioxide and water via cellular respiration. However, under anaerobic conditions yeast cannot use the end products of glycolysis to generate energy in cellular respiration. Instead, they rely on a process called fermentation[24]. Fermentation converts pyruvate into ethanol through the intermediate acetaldehyde.

Water

Water can often play a very important role in the way a beer tastes, as it is the main ingredient. The ion varieties present in water can affect the metabolic pathways of yeast, and thus the metabolites one can taste. For example, calcium and iron are essential in small amounts for yeast to survive, because these metals are usually required cofactors for yeast enzymes.

Carbonation

In aerobic conditions, yeast turns sugars into pyruvate then converts pyruvate into water and carbon dioxide. This process can carbonate beers. In commercial production, the yeast works in anaerobic conditions to convert pyruvate into ethanol, and does not carbonate beer. Beer is carbonated with pressurized CO_2. When beer is poured, carbon dioxide that is dissolved in the beer forms bubbles. These bubbles grow and accelerate as they rise by feeding off of nearby smaller bubbles, a phenomenon known as Ostwald ripening. These larger bubbles lead to "coarser" foam on top of poured beer.

Nitro beer

Beers can be carbonated with CO_2 or with other gases such as Nitrogen. These gases are not as soluble in water as carbon dioxide, so they form bubbles that do not grow through Ostwald ripening. This means that the beer has smaller bubbles and a more creamy and stable head. This less soluble gas gives the beer a different and flatter texture. In beer terms, the mouthfeel is smooth, not bubbly like beers with normal carbonation. Nitro beer could taste less acidic than normal beer.

Soft Drinks

A soft drink is a drink that usually contains carbonated water (although some lemonades are not carbonated), a sweetener, and a natural or artificial flavoring. The sweetener may be a sugar, high-fructose corn syrup, fruit juice, a sugar substitute (in the case of diet drinks), or some combination of these. Soft drinks may also contain caffeine, colorings, preservatives, and/or other ingredients.

One of the most heavily promoted and "coolest" products around are soft drinks. Millions are spent on superstars advertising various soft drinks. We know they are not healthy but how damaging is that innocent - and many would say satisfying - can of coke to our health? One can of soft drink has about:

- 10-13 teaspoons of sugar
- 30 to 55 mg of caffeine
- And additionally it is loaded with artificial food colors and sulphites.

Physical costs from soft drink ingredients:

- **Sugar:** Soft drink manufacturers are the largest single user of refined sugar in the United States. It is a proven fact that sugar increases insulin levels, which can lead to high blood pressure, high cholesterol, heart disease, diabetes, obesity, premature aging and numerous other negative side effects including associated tooth disease problems.
- **Caffeine:** Caffeinated drinks cause many nervous system reactions including the jitters and insomnia. High caffeine use is associated with high blood

pressure, irregular heartbeat, elevated blood cholesterol levels, vitamin and mineral depletion that potentially leads to disease.

- ❖ **Aspartame:** This chemical is used as a sugar substitute in diet soda. There are over 92 different health side effects associated with aspartame consumption including brain tumors, birth defects, diabetes, emotional disorders and epilepsy/seizures. Further, when aspartame is stored for long periods of time or kept in warm areas it changes to methanol, an alcohol that converts to formaldehyde and formic acid, which are known carcinogens (cancer causing agents).

- ❖ **Phosphoric Acid:** May interfere with the body's ability to use calcium, which can lead to osteoporosis or softening of the teeth and bones. Phosphoric acid also neutralizes the hydrochloric acid in your stomach, which can interfere with digestion, making it difficult to utilize nutrients.

- ❖ **Benzene:** Recent random government testing of soft drinks found five of them contaminated with levels of benzene (which is a cancer-causing chemical linked to leukemia) that exceeded federal standards set for benzene in drinking water. Benzene is able to form in beverages that contain vitamin C (ascorbic acid) and the preservatives sodium benzoate or potassium benzoate. Exposure to heat and light can trigger the formation.

Soft drinks are one of the main reasons from a nutritionally perspective why many people suffer health problems. Aside from the negative effects of the soda itself, drinking a lot of it is likely to leave you with little appetite for whole foods that your body needs to function at its best.

The Difference between Energy Drinks and Soft Drinks

To many of us, the energy drinks seem to little more than a 'specialized type of soft drinks.' It is this attitude that drives many of us to use the energy supplement in pretty much the same way we use soft drinks which is casually.

But the energy drinks are a different 'kettle of fish' altogether.

Now the main thing that makes the difference between energy drinks and soft drinks is their respective potency[25] levels. More specifically, the main difference here is that the energy supplement are much more potent formulations than the soft drinks.

Soft drinks, in terms of formulation, tend to be nothing more than combination of carbonated water, sugar and food coloring, with a preservative here and there to make soda. In the case of a juice, which may qualify to be termed as a soft drink in some circles; the only difference we may be looking at is where the water used is not carbonated and where some fruit extract is added for the sake of authenticity. That is just about it all.

In energy supplement, however, we tend to be looking at much more potent substances than water, fruit extract, sugar and preservative. In the most basic of energy drinks, we will tend to be looking at copious levels of caffeine, from which they obtain their stimulant effect. Now in smaller quantities, caffeine may be an absolutely harmless substance, but in the amounts that tend to be employed in these energy supplement, it becomes truly potent and capable of bringing about remarkable changes in the body. The amounts of caffeine employed in energy drinks will tend to be the sorts of amounts that that can cause real euphoria[26] when taken in excess. These would also be the sorts of amounts that can results in seizures, as their effects wear off from the body forcing the body to go back to an exponentially lower energy level.

The natural energy supplement are no less potent. In order to make them capable of energizing the body, these drinks will tend to contain substances such as Ginkgo Biloba, Ginseng or the acai berry extract which turn out to be extremely potent substances. It is through the action of these substances that the energy supplement are able to accelerate user's metabolism rates to bring about the energizing effects.

Once one gets to understand this difference that exists between energy drinks and soft drinks, they will tend to treat the energy drinks with the seriousness they deserve. That would be by among other things, keeping them away from the reach of children, using them in the right amounts and reading the instructions for their usage keenly. It would also mean avoiding using them alongside things they are likely adversely interact with, such as alcohol.

CHAPTER 5

Chemistry in Toiletries

Introduction

A recent news item claimed that using make-up and other toiletries on a daily basis can mean up to 5 pounds of chemicals being absorbed into your body in a year. We really need to start questioning the products we are putting on our skin and not just assume that the chemicals in them are safe. Do you know that absorbing chemicals through the skin could be more dangerous than swallowing them - whereas lipstick getting into your mouth is broken down by the enzymes in saliva and in the stomach, chemicals in products which are absorbed through the skin get straight into your bloodstream.

The US National Institute of Occupational Safety and Health has reported that nearly 900 of the chemicals used in cosmetics are toxic, although other groups believe that the real number is much higher. The chemicals most often cited as giving cause for concern include:

- parabens, preservatives used in products including soap, shampoo, deodorant and baby lotion which may cause endocrine (hormonal) effects.
- sodium lauryl sulphate, used to help create lather in soaps, shampoo, shaving foam, toothpaste and bubble bath, which can cause skin irritation.
- phenylenediamine, an ingredient in hair dyes which is thought to be carcinogenic (cancer-producing).
- formaldehyde, a preservative used in aqueous products such as shampoo, conditioner, shower gel, liquid hand wash and bubble bath. It is claimed to be a carcinogen and sensitiser.
- phthalates, common in fragrances and nail polish, which are thought to cause reproductive or developmental effects.

Using cosmetics may not just be a risk to the person who uses them. For instance, the ingredient nonoxynol (or nonylphenol ethoxylate): breaks down in water treatment into nonylphenol, a synthetic estrogen that feminises male fish, and triclosan, an antibacterial widely used in soaps, may be contributing to the general problem of antibiotic resistance

When we think about the damage that chemicals may be doing to our health, we tend to focus on our food and drink, and the air we breathe. It is easy to forget that the cosmetics and toiletries we put on our skin every day may be equally responsible for polluting our bodies and the environment.

What can you do ?

Your first step in creating a fresh, natural bathroom is to get rid of all the old stuff. Haul out all those dusty old skin

Chemistry in Toiletries

care products you opened but never finished as they will have deteriorated over time and most should be used within either 6 or 12 months of opening.

Once you have sorted through all the old, it is time to deal with the products you might have in there which may not be so good for your skin or your body. Get the magnifying glass out - it is time to look through the listed ingredients!

Considering throwing away products which have any of the following ingredients in:

- Sodium Lauryl Sulphate - SLS can dry your skin out, cause contact eczema and irritate your eyes and gums.
- Parabens (normally listed as -paraben, such as methylparaben) - Some parabens have been shown in lab tests to have a mild hormone-like effect.
- Propylene glycol - This chemical can transport other ingredients through the skin and into the body.
- Ethyl alcohol - Ethyl alcohol is drying on the skin and many people react to it.
- Lanolin – Cosmetic lanolin comes from sheep wool, which has often come into contact with the harsh chemicals found in sheep dip.

You should check everything from your skin care products to your tooth paste, and your liquid soap to your shampoo, as they can all contain nasties. It might be hard to say goodbye to those familiar pots and potions but don't worry, you'll feel better!

Natural Vs Chemical Toiletries

Why should you choose toiletries made with natural ingredients over those with added chemicals?

Natural ingredients are usually softer, gentler & therefore kinder & more caring to your skin & hair. Have you ever noticed how your skin feels 'dry' after showering & bathing? This can be caused by a variety of chemical additives in your shower gels, soaps or body washes which as part of the cleansing action remove 'sebum', our body's natural skin & hair lubrication.

Take a look at the ingredients listed on the products you use to see if they contain DEA[27], MEA[28] or TEA[29]; these can be the cause of that 'dry' feeling. Natural based toiletries often use Aloe Vera, Vegetable Glycerin, a variety of plant oils & even honey to help your skin & hair to retain & replenish moisture to keep your skin soft & supple.

Golden Jojoba oil (simmondsia chinensis) is a popular plant oil & is often found in many skin & hair preparations as this oil is the closest in composition to Sebum & is extremely gentle & compatible with our skin & hair.

Do natural toiletries cost more than chemical based ones?

Initially this may seem to be so BUT it is quite common when using natural products that you only need to use a fraction of the amount of product compared to a chemical based one. Therefore ml to ml a natural product can last a lot longer & be more cost effective.

Those who suffer with skin sensitivities &/or allergies may also notice that their skin becomes less irritated & even becomes softer & smoother. Chemical based or natural - which will you choose?

But what are essential oils?

Essential oils are highly concentrated aromatic plant extracts which have been used throughout history for their therapeutic abilities to relax & unwind, revive & uplift, soothe & relax the body, mind & spirit. These oils are extracted by steam distillation[30] or expression from the roots, stems, leaves, petals, fruits or nuts of the plant.

Essential oils should be diluted before applying to the skin. Just a couple of drops added to body lotion or a carrier oil is all that is required to be able to benefit from the therapeutic properties of these natural oils.

These oils can affect our mood, emotions, health & well-being. Pain relief, hormone balancing, skin soothing, stress relieving, sleep inducing are but a few of the properties these wonderful oils possess & they can also help to combat bacterial, viral & fungal infections.

What are the Benefits of Handmade Soaps and Toiletries?

Handmade soaps and toiletries are extremely beneficial to the average consumer. They are well made and hand crafted body products that are complimentary to your skin and they contain no synthetics of any kind. There are no chemicals that can be detrimental to your body. When was the last time you visited your local drugstore? When was the last time you actually read the labels on the soaps that you bought? Do you even know or understand half the words in the ingredient list? Handmade soaps are handcrafted with care so there are no synthetic products of any kind. Small batches are created to ensure the quality of the product and they are not tested on animals.

Handmade soaps are unbelievably rich and moisturizing. They soothe and soften the skin as well as cleanse the skin without drying it out. They are made out of renewable, natural, botanical, and vegetable based resources. They are gentle enough even for the most sensitive and allergy prone skin. Most producers of handmade products only use the best of ingredients such as olive oil or shea butter.

They add wonderful and pure essential oils that are extracted from plants to act as therapeutic agents to the skin and to make it smell great. Only the best and most natural ingredients are used in developing handmade soaps to ensure benefits for you and your body.

Handmade soaps and toiletries also benefit the environment. For example using natural products that don't contain any synthetic materials won't clog or ruin pipes. Also, the remains that go through our pipes end up in our waters. So another great benefit of handmade soaps is that they don't pollute as bad because there are no chemicals such as in the average detergent bars. We should all do our part to help out.

CHAPTER 6
Chemistry in Agriculture

Importance of Chemicals in increasing agricultural yield

The importance of chemicals in increasing the agriculture production can't be set aside. If we go through the history of India we find an enormous rise in the wheat production during 1970s decade, which heralded green revolution in the country. The green revolution was impossible without compost fertilizers, pesticides, germicides, etc. When green revolution took place, the country was facing the world's worst recorded food disaster continued from 1943 in British-ruled India. At that time, chemical industry played a significant role in increasing the production of agriculture products by providing scientifically developed fertilizers, pesticides and germicides. Prior to this, Indian agriculture sector was facing worst effects of insects and low productivity. Today India comes in the list of eminent agricultural products exporters that comprises of basmati rice, cotton, spices, tobacco, etc. This is possible just because of best Fertilizers, Pesticides and Germicides.

Farmer spraying Pesticides in Rice Fields

What are Pheromones?

Organisms release chemicals, which help them to communicate with other members of the species, into the environment. These chemical substances are known as Pheromones. The extensive use of pheromones is visible in the day-to-day activities of insects. The most widely studied pheromones are those released by the female moths, which attract specific males for mating.

Pheromones are substances which are volatile and odorous. These cause physiological reactions like modulating sexual activity, bringing out aggression, and territory making. There are different types of pheromones for stimulating different reactions. These organisms use trail pheromones to move from nests to their food and back.

Pheromones act as sex-attractants which are usually emitted by females and can attract males from over two miles away. Although, there are some male insects which also produce them. A small amount of sex attractant of an insect pest can collect all the males in the vicinity and they can be disposed off or sterilized. It is highly specific method because each insect has its own attractant. There is no spraying , hence no pesticide residue. In addition, the concentration is so small that it does not affect any other species. For example, the gypsy moth attractants attract male moths in the areas when a trap is baited with 1×10^{-9}g. Gypsy moths are highly voracious eaters and will completely denude trees if they go unchecked.

Chemistry in Agriculture

Why is Microbial Fertilizer the Best Option for Agriculture?

Microbial fertilizer has cemented itself as a popular option to chemical fertilizers throughout the agriculture industry. Microbial fertilizer utilizes living microorganisms to increase the supply of primary nutrients to plants. Depending upon the type of microorganisms included in the fertilizer, primary nutrients are provided via a variety of natural processes including solubilizing phosphorus, synthesizing plant-specific growth-promoting substances and nitrogen fixation. For example, certain microbes have a capacity to utilize naturally occurring phosphorous to create the solubilized precipitated phosphates needed by plants. The result is a symbiotic relationship in which the bacteria provides soluble phosphate and plants supply root born carbon compounds needed to sustain bacteria growth.

Chemical Fertilizers Lead to a Long-Term Decline in Productivity:

Today, industry experts note that fields which rely solely on chemical fertilizer applications are experiencing a slow decline in long term productivity. This is especially true for monoculture fields. A primary reason chemical fertilizers lead to a decline in productivity is the soil becomes oversaturated with the chemicals being applied to it. Over time, this limits biodiversity within the soil and undercuts helpful, natural nutrient producing processes.

Microbial Fertilizers Directly Improves Soil Strength and Nutrient Density:

Microbial fertilizers outperform the chemical alternatives in a variety of ways. One of the most notable benefits is improved soil strength and nutrient density. This type of fertilizer, once applied, will increase soil strength by activating effective microbes such as non-sulfur bacteria. A type of purple non-sulfur bacteria known as Rhodobacter sphaeroides is rich in nutrients and has the ability to create key plant hormones. As a result, this bacterium creates a beneficial symbiotic relationship with the plants being grown.

Microbial Fertilizers Can Reduce the Need of Certain Pesticides:

A common drawback of chemical fertilizers is they must be used in conjunction with potentially harmful pesticides in order to maintain a robust crop. This creates a significant expense in terms of product cost, labor, and time. By opting for microbial fertilizer the likelihood, an additional pesticide will be needed is dramatically reduced. Numerous studies have shown that microbial fertilizers not only increase crop output but can also significantly reduce or eliminate the need for certain pesticides. This is possible because the microbes inhibit and prevent the propagation of many common agricultural pests.

Microbial Fertilizers vs. Chemical Fertilizers:

Overall, microbial fertilizers offer numerous benefits over their chemical counterparts. Not only does it offer the potential to increase crop yields and reduce the need for pesticides; but it also eliminates the risk of a long-term decline in productivity associated with extended use of chemical fertilizers. Agriculture soil is a constantly evolving system of plants and microbes. The ability to leverage the natural, symbiotic relationship of these two elements makes microbial the best option.

Using Organic Agriculture Products

The Organic Food Market has grown quickly since 1990. In just 12 short years it reached $63 billion dollars and that number is still climbing.

Organic gardening is quite simple and very basic. It's not as complicated as many people think it to be. Organic methods of gardening include using microorganisms that will suppress weed growth while promoting germination of the plants.

High density planting is another method that is frequently used and it works well. By planting tight close rows there is no room for weeds to grow. This helps to encourage rapid germination as well. Waiting to plant until the soil is warmer is also a good way to get started.

There are many ways to mulch and use plastic as well to eliminate weed growth. Mulching[31] is great as it helps to hold the water in while blocking the weed growth through the mulch.

When it comes to insects there are also some great ways to combat the little critters. There are traps and sound set ups that will zap the predatory insects or one can simply invest in some beneficial insects such as pirate bugs or lady bugs to eat the pests. Using organic agriculture products is an ideal way to ensure that the chemicals used in commercial fertilizers never touch your dinner plate.

The commercial fertilizers can have added compounds that are known to irritate certain people. In some people some of these chemicals can cause cancers and other reactions so many people today are leaning more toward organic food in lieu of their chemically enhanced counter products.

There are some naturally derived organic insecticides that can also be used. These include Bacillus thuringiensis and a chrysanthemum extract. However, less than 10 percent of organic farmers choose this route and most will prefer to use such methods as lady bugs to combat their bugs.

Composting is another method that works well for organic gardening. Recycling fruit and vegetable waste that would have been thrown out and adding in other organic matter such as lawn clippings, leaves, newspaper and the like can create a delightful compost full of rich nutrients.

Red wiggler worms then come in and eat up the parts of the compost and excrete what they don't need. This mixed in with garden soil is an ideal way to fertilize plants and encourage good growth.

Organic meats are from animals that are raised in organic pastures. Organic farming is simply taking advantage of simple methods of encouraging plant growth. They avoid the use of anything that has been genetically engineered.

There are no GMOs in organic farming and many organic farmers worry about their crops being pollinated by bees that have been to farms that have chemicals on their crops. Regulations vary on how to combat this but do include crop rotations and leaving some fields fallow for a season to avoid this cross contamination.

Biopesticides are Likely to Gain Momentum

Biopesticides are showing large-scale adoption by the industry, due to less regulation on product approval and low costs of product development. Being essentially less harmful, when compared to regular synthetic pesticides, biopesticides typically affect only the target pest and other organisms that are closely related to it, as against broad-spectrum conventional pesticides, which can be harmful to other organisms. It takes nearly USD 250 million and 10 years to develop a new pesticide product; for the development of GM crop, it takes 12-13 years and roughly USD 130 million. However, a biopesticide can come to market in 3-5 years, with roughly USD 3-5 million developmental cost.

A key factor is likely to be greater R&D investment in the area, now that many of the major agrochemical companies have an interest in the sector. This, coupled with the market opportunities listed above, suggests that the biopesticide sector may perform ahead of the crop protection sector as a whole. The trend of the market can be seen to be changing from a single-component formulation to a multi-component formulation, making it more suitable for two or more pests.

CHAPTER 7
Chemistry involvement in The construction industry

Concrete building under construction

The construction industry has undergone rapid and drastic changes in the last few decades. The steady development of science and technology has made the issue of construction easier and less time consuming. Today most of the energy of the industry goes into

Chemistry involvement in The construction industry

designing and building superstructures like skyscrapers, dams and bridges. These structures are not built only for a few days, instead they are designed to last long. Therefore, it is important to build them in such a way that would help them stand tall for years that also without much maintenance work. This is why the construction industry nowadays make use of certain chemicals that serve the desired purpose quite efficiently.

The Chemicals used in Construction

There are different types of chemicals that are used while building a structure. These chemicals differ from each other in terms of properties and the functions that they perform. Their method of application also depends on the purpose that they serve. Some chemicals are mixed with the concrete while others are coated on the concrete blocks after they have been installed in their designated places. One of the most widely used chemicals in construction is the waterproofing chemical. It is used to protect the reinforced concrete from the erosion caused by water over years. It is generally applied to the outside wall and the roof, though it can also be used on the inner walls, ceiling and floor.

The chemicals that are used in construction projects are with a special purpose and that is to give the structure stability. Some of the chemicals which are used in the construction more commonly are concrete admixture, waterproofing, surface treatment, grout, floor topping, floor hardening, coating, and sealer. Especially, silicone sealant and concrete admixture can strengthen the structures and protect them from severe weather, ground shifting and other environment hazards. And therefore, construction chemicals are important elements of quality construction materials.

Constructing Roofs and Walls

Bricks, concrete and cement are among the most important building materials that are used for constructing roofs and walls. In addition, these constructions are strengthened by adding materials like iron and steel. These two materials ensure long life of the construction and make building capable of resisting damage caused due to different factors. Pillars are other parts of building constructions that make use of these materials to come out as strong structures. It is also worth mentioning that paints and emulsions should be counted among the materials that take part in completing these constructions.

Constructing Floors

Earlier, bricks and cement were used for the construction of floors as well. However, the wonderful breed of natural stones and tiles came into being and it laid the foundation for beautiful and astonishing flooring. Marble, granite, limestone and a variety of other materials are in use nowadays for laying enchanting floorings. Going further, these natural materials are also in use for constructing staircases and countertops for kitchens and other areas of different types of buildings. Outdoor areas, driveways and swimming pools too make use of these aesthetically appealing building materials.

Constructing Doors and Windows

A building is incomplete until and unless it has provisions for doors and windows. Needless to say, different varieties of wood have been the most popular materials for building these parts. Mahogany, oak, maple and many other wood varieties make arrangements for these constructions. In addition, glass is a widely used material for building doors and windows. The material is also available in different varieties to meet different tastes and preferences. Iron and steel are used for building products like grills and scrolls that are attached to windows.

For Bathroom Fittings

In addition to the building materials discussed above, another category is that of materials used for plumbing and bathroom fittings. Stainless steel, chromium, nickel, brass and many such materials are available for different styles of bathroom fittings. Porcelain and ceramic are other two materials that have gained recognition in this direction. As far as plumbing requirements are concerned different varieties of plastics are useful for the same. PVC, for example, is one of the popular choices among the building specialists.

Chemicals used as pest control in construction of new house

There are many chemicals available in markets that are used as pre-treatment chemicals while construction of a new house. These products guarantee protection of your house for a long period of time from different pests. The Best chemical used for protecting wood from attack of termites is termidor. This is a very effective chemical. It stops termites from doing any damage to your walls of wood. It is not a repellent. It almost kills termites as they enter the treated area. There are many insecticides in this chemical. When these termites start biting the wood, this chemical goes into their stomach and causes death.

Another very good termiticide that can be used is Permethrin 36.8%. This is an effective chemical that stays

Chemistry involvement in The construction industry

longer on walls and makes them safe. It is good for ants, beetles, bag-worms, fleas, flies, mosquitoes, ground beetles and many more. You can apply this chemical on interior side of walls, on landscape gardens including lawns, parks and ground, tiles and other building material.

Demon TC[32] and Cyper TC[33] are used as Pest control ingredients. These products not only kill the termites but also repel other termites. This chemical will last for almost 6 years. You can use it under slabs as pre-construction material. You can use as post construction material too. You can use them along walls, slabs and other places that can be infected. These chemicals can also be used for killing other pests other than termites.

Dominion 2L or Premise 2 Termiticide are very good and well concentrated chemicals that are effective for treating wood.

CHAPTER 8

Various Industrial Chemicals and Their Uses

Man working with industrial chemicals

Various types of industrial chemicals are manufactured by a large number of manufacturers for industrial applications. Today these chemicals are used for different purposes including the manufacture of explosives, different types of fuels used in engines,

Various Industrial Chemicals and Their Uses

various types of dyes and paints, insulators, cosmetics, detergents and soaps.

The chemical industry has grown rapidly in the past several decades. Industrial chemicals are used in various sectors of the economy including agriculture, service sector, construction sector and manufacture of consumer goods. The use of chemicals is wide and varied with a large number of consumers all across the world.

Various Types of Industrial Chemicals

There are more than hundreds of different varieties of chemicals produced each year for various manufacturing purposes. Some of the most significant types of chemicals include the following:-

- ❖ **Chlorine:** One of the most widely used chemicals in the manufacture of bleaching agents includes chlorine. But now chlorine is replaced by the chemical called hydrogen peroxide as chlorine is considered to be harmful to the environment.

- ❖ **Phosphoric acid:** Industrial chemicals also include phosphoric acid which is used for the production of fertilizers. It is also used in manufacture of food products and soft drinks.

- ❖ **Nitrogen:** Nitrogen is an inert substance that is often used in industries to control and avoid thermal reactions. It is also used in solid conveying gas carrier as a substitute for air.

- ❖ **Sodium carbonate:** One of the most commonly used chemicals includes sodium carbonate or soda ash. It is used in the manufacture of glass and different types of cleaning agents.

- ❖ **Sulfuric acid:** It is one of the most widely used chemical acid which helps to neutralize alkaline substances.

- ❖ **Sodium silicate:** It is one of the most commonly used chemical for industrial insulation.

- ❖ **Aluminum sulfate:** It is used in waste water treatment plants and in the manufacture of paper.

- ❖ **Sodium hydroxide:** Sodium hydroxide is widely used in the manufacture of soaps and dyes. It is one of the commonly used alkaline substances in industries. As it has the ability to neutralize acids, it can be used as a good cleaning agent.

- ❖ **Ammonium nitrate:** This white crystal-like substance is widely used as an agricultural fertilizer. It is also used in cold packs to reduce injuries and swellings.
- ❖ **Urea:** Another important industrial chemical is urea which is mainly used to fertilize agricultural land. Various types of livestock feed also contain urea.
- ❖ **Hydrochloric acid:** One of the most useful chemicals includes hydrochloric acid which is largely used in pharmaceutical industries. It is also used to produce other chemicals.
- ❖ **Potash:** Another commonly used chemical for agricultural purposes includes potash. It is also used to manufacture soaps, glass and ceramics.
- ❖ **Titanium dioxide:** This chemical is a white pigment that is used in the manufacture of various products ranging from paints to food items. It is also used in various pharmaceutical products.
- ❖ Other common industrial chemicals include acetone, carbon black, acetic acid, propylene oxide, ethylene oxide, methanol, vinyl chloride and so forth.

CHAPTER 9

Chemistry in Rocket Propellants/Space Vehicles

Rocket propellant is the reaction mass[34] of a rocket. This reaction mass is ejected at the highest achievable velocity from a rocket engine to produce thrust[35]. The energy required can either come from the propellants themselves, as with a chemical rocket, or from an external source, as with ion engines.

Rockets create thrust by expelling mass backward at high velocity. The thrust produced can be calculated by multiplying the mass flow rate of the propellants by their exhaust velocity relative to the rocket (specific impulse). A rocket can be thought of as being accelerated by the pressure of the combusting gases against the combustion chamber and nozzle, not by "pushing" against the air behind or below it. Rocket engines perform best in outer space because of the lack of air pressure on the outside of the engine. In space it is also possible to fit a longer nozzle without suffering from flow separation.

Most chemical propellants release energy through redox chemistry, more specifically combustion. As such, both an

Chemistry in Rocket Propellants/Space Vehicles

oxidizing agent[36] and a reducing agent[37] must be present in the fuel mixture. Decomposition, such as that of highly unstable peroxide bonds in monopropellant rockets, can also be the source of energy.

In the case of bipropellant liquid rockets, a mixture of reducing fuel and oxidizing oxidizer is introduced into a combustion chamber, typically using a turbo pump to overcome the pressure. As combustion takes place, the liquid propellant mass is converted into a huge volume of gas at high temperature and pressure. This exhaust stream is ejected from the engine nozzle at high velocity, creating an opposing force that propels the rocket forward in accordance with Newton's laws of motion.

Chemical rockets can be grouped by phase. *Solid rockets* use propellant in the solid phase, *liquid fuel* rockets use propellant in the liquid phase, *gas fuel* rockets use propellant in the gas phase, and hybrid rockets use a combination of solid and liquid or gaseous propellants.

In the case of solid rocket motors, the fuel and oxidizer are combined when the motor is cast. Propellant combustion occurs inside the motor casing, which must contain the pressures developed. Solid rockets are typically have higher thrust, less specific impulse, shorter burn times, and a higher mass than liquid rockets, and additionally cannot be stopped once lit.

Most liquid chemical rockets use two separate propellants: a fuel and an oxidizer. Typical fuels include kerosene, alcohol, hydrazine and its derivatives, and liquid hydrogen. Many others have been tested and used. Oxidizers include nitric acid, nitrogen tetroxide, liquid oxygen, and liquid fluorine.

Materials used in Making International Space Station

International space station (ISS)

Building a home for living in space requires a little more than plywood and two-by-fours. Titanium, Kevlar, and high-grade steel are common materials in the ISS. Engineers had to use these materials to make the structure lightweight yet strong and puncture-resistant.

Because each of the aluminum-can shaped components of the Station has to be lifted into orbit, minimizing weight is crucial. Lightweight aluminum, rather than steel, comprises most of the outer shell for the modules.

To ensure the safety of the crew, the Space Station wears a "bullet-proof vest." Layers of Kevlar, ceramic fabrics, and other advanced materials form a blanket up to 10 cm thick around each module's aluminum shell. (Kevlar is the material used in the bullet-proof vests used by police officers.)

Glass used in Spacecrafts

The inner glass pane made of tempered alumino-silicate glass is called pressure glass. It is designed to resist as much as possible the pressure of the shuttle cabin in the vacuum of space.

CHAPTER 10

Chemistry Involved in Clothing

Clothing is often made of cloth. There are many different types of cloth, with different names and uses. Main differences between types of cloth include how the cloth is made (woven, knitted, felted, and how those techniques were implemented), what fiber it is made from, and what weight the cloth is. Different types of cloth may be used for different types of clothing.

Examples of clothing materials

Common natural clothing materials are:

- Fabric made of cotton, flax, wool, silk
- Denim
- Leather
- Fur

Other materials are made from synthetic fibers, primarily from petrochemicals[38], which are not generally biodegradable. Common synthetic materials include:

Chemistry Involved in Clothing

- ❖ NYLON

Nylon Jacket

- ❖ Nylon was first produced in 1935. Nylon is a thermoplastic[39] silky material. It became famous when used in women›s stockings ("nylons") in 1940. It was intended to be a synthetic replacement for silk and substituted for it in many different products .

- ❖ POLYESTER

Polyester scarf

Polyesters include naturally occurring chemicals and synthetics. Natural polyesters and a few synthetic ones are biodegradable, but most synthetic polyesters are not. Polyesters may change shape after the application of heat and are combustible at high temperatures. They tend to shrink away from flames and self-extinguish upon ignition. Polyester fibers have high tenacity[40] and E-modulus[41] as well as low water absorption and minimal shrinkage in comparison with other industrial fibers.

- ❖ Spandex (elastane) is known for its exceptional elasticity. It is stronger and more durable than rubber, its major non-synthetic competitor.

Some less common clothing materials are:

- ❖ Acetate
- ❖ Cupro
- ❖ Flannel
- ❖ Lyocell
- ❖ PVC-Polyvinyl chloride
- ❖ Rayon
- ❖ Recycled or Recovered Cotton
- ❖ Other Natural Fibers, including bamboo, jute, and hemp

Reinforcing materials such as wood, bone, plastic and metal may be used in fasteners or to stiffen garments.

Why Is Natural Organic Clothing Vital For Your Health?

Organic clothing is a subject worth learning about. Just like food, clothing affects us intimately in different ways:

1. Through our personal well-being and health
2. Through the effects on our environment and well-being of workers

Our skin being the largest organ is entrusted with protecting us from dangers of the outside world. Clothes become the protector of the skin in turn but there are problems with that. 70% of clothes sold now are "synthetic". What that means, is that it is derived straight from crude oil, with the aid of heavy chemicals, just like making plastic. The resulting textiles offer no ventilation and no cooling, so with body heat and moisture, skin becomes a breeding ground for bacteria. Any toxins sweated out become re-introduced and re-absorbed again.

It is only the last 60 years that chemistry got so much involved in our clothes. Historically clothes were made of natural textiles, mostly plant based - cotton, hemp, linen or animal source like wool or silk. Many of the natural fibers have been around for hundreds of thousands years. Our bodies become friends with them. Each textile has unique qualities and offer usually good ventilation, allowing for regulation of skin temperature.

"Organic clothing", akin to organic food - means clothes that are made from fibers grown to specific organic standards, like Global Organic Textile Standards. Most common is organic cotton and there is a good reason to support organic cotton. Conventional (non-organic) cotton depends heavily on pesticide and insecticide, up to 25% of world use just for cotton. These chemicals cause headaches, fatigue, nausea but also cancers, neurological disorders and birth defects.

Textile manufacturing employs also a slew of harmful chemicals. As shown in series of extensive reports commissioned by Greenpeace, many garments contain NPEs (nonylphenol ethoxylates). The insidious effect of that chemical is mistaken by the body as hormone, like estrogen and causes hormonal havoc. Even fish exposed to NPEs become part male, part female.

The toxic chemicals in our lives have already seeped into our sofas, carpets and many other products. Children sleep in flame retardant soaked pajamas. We are in true toxic overload. There are solutions though. Our bodies can handle a lot of problems, just care needs to be taken what you put in it and on it. NO nasty sprays, no pesticides, no formaldehyde. Each of us might be sensitive to different item, so to stay safe, take one step at the time with basics layers that are next to your skin. Look for comfort, ventilation and simplicity and look for organic designation of the textile.

Different Dyes used to the Fabrics

Dyeing is the application of dyes or pigments on textile materials such as fibers, yarns, and fabrics with the goal of achieving color with desired color fastness. Dyeing is normally done in a special solution containing dyes and particular chemical material. Dye molecules are fixed to the fibre by absorption, diffusion[42], or bonding with temperature and time being key controlling factors. The bond between dye molecule and fibre may be strong or weak, depending on the dye used.

The primary source of dye, historically, has been nature, with the dyes being extracted from animals or plants. Since the mid-19th century, however, humans have produced artificial dyes to achieve a broader range of colors and to render the dyes more stable to washing and general use.

Fibre content determines the type of dye required for a fabric:

- ❖ Cellulose fibres: cotton, linen, hemp, ramie, bamboo, rayon
- ❖ Protein fibres: wool, angora, mohair, cashmere, silk, soy, leather, suede

Cellulose fibres require fibre-reactive, direct/substantive, and vat dyes, which are colourless, soluble dyes fixed by light and/or oxygen. Protein fibres require vat, acid, or indirect/mordant dyes, that require a bonding agent. Each synthetic fibre requires its own dyeing method, for example, nylon requires acid, disperse and pigment dyes, rayon acetate requires disperse dyes, and so on. The types of natural dyes currently in use by the global fashion industry include:

Animal-derived dyes

Cochineal insect (red), Cow urine (Indian yellow), Lac insect (red, violet), Murex snail (purple), Octopus/Cuttlefish (sepia brown) etc.

Plant-derived dyes

Cutch tree (brown), Gamboge tree resin (dark mustard yellow), Chestnut hulls (peach to brown), Himalayan rhubarb root (yellow), Indigofera plant (blue), Kamala tree (red), Larkspur plant (yellow), Madder root (red, pink, orange), Mangosteen peel (green, brown, dark brown, purple, crimson), Myrabolan fruit (yellow, green, black), Pomegranate peel (yellow), Teak leaf (crimson to maroon) etc.

Types

Dyes are classified according to their solubility and chemical properties.

Acid dyes are water-soluble anionic dyes that are applied to fibers such as silk, wool, nylon and modified acrylic fibers using neutral to acid dye baths[43]. Attachment to the fiber is attributed, at least partly, to salt formation between anionic groups in the dyes and cationic groups in the fiber. Examples of acid dye are Alizarine Pure Blue B, Acid red 88, etc.

Basic dyes are water-soluble cationic dyes that are mainly applied to acrylic fibers, but find some use for wool and silk. Usually acetic acid is added to the dye bath to help the uptake of the dye onto the fiber. Basic dyes are also used in the coloration of paper.

Direct or substantive dyeing is normally carried out in a neutral or slightly alkaline dye bath, at or near boiling point, with the addition of either sodium chloride (NaCl) or sodium sulfate (Na_2SO_4) or sodium carbonate (Na_2CO_3). Direct dyes are used on cotton, paper, leather, wool, silk and nylon. They are also used as pH indicators and as biological stains.

Mordant dyes require a mordant, which improves the fastness of the dye against water, light and perspiration. The choice of mordant is very important as different mordants can change the final color significantly. Most natural dyes are mordant dyes. The most important mordant dyes are the synthetic mordant dyes, or chrome dyes, used for wool; these comprise some 30% of dyes used for wool, and are especially useful for black and navy shades. The mordant potassium dichromate is applied as an after-treatment. It is important to note that many mordants, particularly those in the heavy metal category, can be hazardous to health and extreme care must be taken in using them.

Vat dyes are essentially insoluble in water and incapable of dyeing fibres directly. However, reduction in alkaline liquor produces the water-soluble alkali metal salt of the dye. This form is often colorless, in which case it is referred to as a Leuco dye, and has an affinity for the textile fibre. Subsequent oxidation reforms the original insoluble dye. The color of denim is due to indigo, the original vat dye.

Reactive dyes utilize a chromophore attached to a substituent that is capable of directly reacting with the fiber substrate. The covalent bonds that attach reactive dye to natural fibers make them among the most permanent of dyes. "Cold" reactive dyes, such as Procion MX, Cibacron F, and Drimarene K, are very easy to use because the dye can be applied at room temperature.

Reactive dyes are by far the best choice for dyeing cotton and other cellulose fibers at home or in the art studio.

Disperse dyes were originally developed for the dyeing of cellulose acetate, and are water-insoluble. The dyes are finely ground in the presence of a dispersing agent[44] and sold as a paste, or spray-dried and sold as a powder. Their main use is to dye polyester, but they can also be used to dye nylon, cellulose triacetate, and acrylic fibers. In some cases, a dyeing temperature of 130 °C (266 °F) is required, and a pressurized dyebath is used.

Azoic dyeing is a technique in which an insoluble Azo dye is produced directly onto or within the fiber. This is achieved by treating a fiber with both diazoic and coupling components. With suitable adjustment of dyebath conditions the two components react to produce the required insoluble azo dye. This technique of dyeing is unique, in that the final color is controlled by the choice of the diazoic and coupling components. This method of dyeing cotton is declining in importance due to the toxic nature of the chemicals used.

Sulfur dyes are inexpensive dyes used to dye cotton with dark colors. Dyeing is effected by heating the fabric in a solution of an organic compound, typically a nitrophenol derivative, and sulfide or polysulfide. The organic compound reacts with the sulfide source to form dark colors that adhere to the fabric. Sulfur Black 1, the largest selling dye by volume, does not have a well defined chemical structure.

Some dyes commonly used in Staining:

Basic Dyes	Acidic Dyes
Basic Fuchsin	Acid Fuchsin
Crystal violet	Congo Red

CHAPTER 11

Chemicals used in Paper Industry

Paper chemicals designate a group of chemicals that are used for paper manufacturing, or modify the properties of paper. These chemicals can be used to alter the paper in many ways, including changing its color and brightness, or by increasing its strength and resistance to water.

Pulp and paper mill

Various Chemicals used in different steps of paper making

Pulping

Chemical pulping involves dissolving lignin[45] in order to extract the cellulose from the wood fiber. The different processes of chemical pulping include the Kraft process, which uses caustic soda and sodium sulfide and is the most common; alternatively, the use of sulfurous acid is known as the sulfite process, and soda pulping which is the least ecologically hazardous utilizing sodium hydroxide or anthraquinone.

Caustic soda is added to increase the pH in the pulping process of fibers. The higher pH of the paper-fiber solution causes the fibers to smoothen and swell, which is important for the grinding process of the fibers.

Bleaching

In the production of white paper, the wood pulp is bleached to remove any color from the trace amounts of lignin that was not extracted in the chemical pulping process. There are three predominant methods of bleaching:

1. Elemental chlorine bleaching uses chlorine and hypochlorite.
2. Elemental chlorine-free bleaching is more environmentally friendly since it eliminates the use of hypochlorite and replaces chlorine with chlorine dioxide or sodium chlorate.
3. Totally chlorine-free bleaching utilizes oxygen and hydrogen peroxide. This is the most environmentally

friendly process since it eliminates all chlorinated pollutants.

Sizing

Most paper types must have some water-resistance to maintain a specific writing quality and printability. Until 1980, the typical manner of adding this resistance was by using a rosin in combination with alum. Today, mainly AKD (alkyl ketene dimer) and ASA (alkenyl succinic anhydride) are used.

Strengthening

Wet-strength: Wet-strength additives ensure that paper retains its strength when it gets wet. This is especially important in tissue paper. Chemicals typically used for this purpose include epichlorohydrin, melamine, urea formaldehyde and polyimines. These substances polymerize in the paper and result in the construction of a strengthening network.

Cationic starch: To enhance the paper's strength, cationic starch is added to wet pulp in the manufacturing process. Starch has a similar chemical structure as the cellulose fibre of the pulp, and the surface of both the starch and fibre are negatively charged. By adding cationic (positive charged) starch, the fibre can bind with the starch and thus also increase the interconnections between the fibres. The positively charged portion of the starch is usually formed by quaternary ammonium cations. Quaternary salts that are used include 2.3-epoxy propyl trimethyl ammoniumchloride (EPTAC, also known as or Glytac Quab, GMAC™) and (3-chloro-2-hydroxypropyl) trimethyl ammonium chloride .

Dry-strength: Dry-strength additives, or dry-strengthening agents, are chemicals that improve paper strength in normal conditions. These improve the paper's compression strength, bursting strength, tensile breaking strength, and delamination resistance. Typical chemicals used include cationic starch and polyacrylamide (PAM) derivatives. These substances work by binding fibers, often under the aid of aluminum ions in paper sheet.

Binders

Binders promote the binding of pigment particles between themselves and the coating layer of the paper. Binders are spherical particles less than 1 μm in diameter. Common binders are styrene maleic anhydride copolymer or styrene-acrylate copolymer. Co-binders, or thickeners, are generally water-soluble polymers that influence the paper›s color viscosity, water retention, sizing, and gloss. Some common examples are carboxymethyl cellulose (CMC), cationic and anionic hydroxyethyl cellulose (EHEC), modified starch, and dextrin. Styrene butadiene latex, Styrene acrylic, dextrin, oxidized starch are used in coatings to bind the filler to the paper. Co-binders are natural products such as starch and CMC (Carboxymethyl cellulose), that are used along with the synthetic binders, like styrene acrylic or styrene butadiene. Co-binders are used to reduce the cost of the synthetic binder and improve the water retention

Fillers

Mineral fillers are used to lower the consumption of more expensive binder material or to improve some properties of the paper. China clay, calcium carbonate, titanium dioxide, and talc are common mineral fillers used in paper production.

Retention

A Retention agent is added to bind fillers to the paper. Fillers, such as calcium carbonate, usually have a weak surface charge. The retention agent is a polymer with high cationic, positively charged groups. An additional feature of a retention agent is to accelerate the dewatering in the wire section of the paper machine. Polyethyleneimine and polyacrylamide are examples of chemicals used in this process.

Coating

Pigments

Pigments that absorb in the yellow and red part of the visible spectrum can be added. As the dye absorbs light, the brightness of the paper will decrease, unlike the effect of an optical-brightening agent. To increase whiteness, a combination of pigments and an optical-brightening agent are often used. The most commonly used pigments are blue and violet dyes.

Optical-brightening agent

Optical brightener is used to make paper appear whiter. Optical-brightening agents use fluorescence to absorb invisible radiation from the ultraviolet part of the light spectrum and re-emit the radiation as light in the visible blue range. The optical-brightening agent thus generates blue light that is added to the reflected light. The additional blue light offsets the yellowish tinge that would otherwise exist in the reflected light characteristics. It thus increases the brightness of the material (when the illumination includes ultraviolet radiation).

CHAPTER 12

Polymers

A polymer (*poly-* "many" + *-mer-* "part") is a large molecule, or macromolecule, composed of many repeated subunits. Due to their broad range of properties, both synthetic and natural polymers play essential roles in everyday life. Polymers range from familiar synthetic plastics such as polystyrene to natural biopolymers such as DNA and proteins that are fundamental to biological structure and function. Polymers, both natural and synthetic, are created via polymerization of many small molecules, known as monomers.

Historically, products arising from the linkage of repeating units by covalent chemical bonds have been the primary focus of polymer science; emerging important areas of the science now focus on non-covalent links. Polyisoprene of latex rubber is an example of a natural/biological polymer, and the polystyrene of styrofoam is an example of a synthetic polymer. In biological contexts, essentially all biological macromolecules—i.e., proteins (polyamides), nucleic acids(polynucleotides), and polysaccharides—

are purely polymeric, or are composed in large part of polymeric components—e.g., isoprenylated/lipid-modified glycoproteins, where small lipidic molecules and oligosaccharide modifications occur on the polyamide backbone of the protein.

Polymers are of two types: naturally occurring and synthetic or man made.

Natural polymeric materials such as hemp[46], shellac[47], amber[48], wool, silk and natural rubber have been used for centuries. A variety of other natural polymers exist, such as cellulose, which is the main constituent of wood and paper.

The list of synthetic polymers, roughly in order of worldwide demand, includes polyethylene, polypropylene, polystyrene, polyvinyl chloride, synthetic rubber, phenol formaldehyde resin (or Bakelite), neoprene, nylon, polyacrylonitrile, PVB[49], silicone, and many more.

Most commonly, the continuously linked backbone of a polymer used for the preparation of plastics consists mainly of carbon atoms. A simple example is polyethylene (‹polythene› in British English), whose repeating unit is based on ethylene monomer. Many other structures do exist; for example, elements such as silicon form familiar materials such as silicones, examples being Silly Putty and waterproof plumbing sealant. Oxygen is also commonly present in polymer backbones, such as those of polyethylene glycol, polysaccharides (in glycosidic bonds), and DNA (in phosphodiester bonds).

Some familiar household synthetic polymers include: Nylons in textiles and fabrics, Teflon in non-stick pans, Bakelite for electrical switches, polyvinyl chloride (PVC) in pipes, etc. The common PET bottles are made of a synthetic polymer, polyethylene terephthalate(PET). The plastic kits and covers are mostly made of synthetic polymers like polythene and tires are manufactured from Buna rubbers. However, due to the environmental issues created by these synthetic polymers which are mostly non-biodegradable and often synthesized from petroleum, alternatives like bioplastics are also being considered. They are however expensive when compared to the synthetic polymers.

"Frying Pan with a non stick coating of TEFLON"

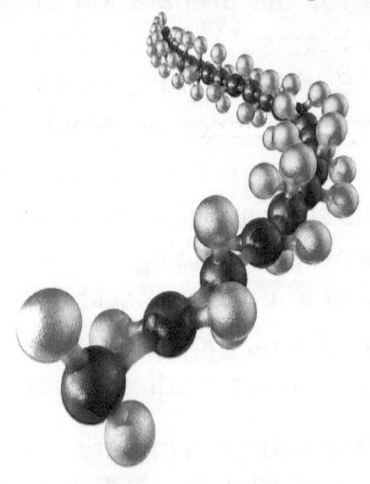

"3D Model of TEFLON molecule"

Organic Polymers

The eight most common types of synthetic organic polymers, which are commonly found in households are:

- Low-density polyethylene (LDPE)
- High-density polyethylene (HDPE)
- Polypropylene (PP)
- Polyvinyl chloride (pvc)
- Polystyrene (PS)
- Nylon, nylon 6, nylon 6,6
- Teflon (Polytetrafluoroethylene)
- Thermoplastic polyurethanes (TPU)]

List of some addition polymers and their uses

Polymer	Abbreviation	Properties	Uses
Low-density polyethylene	LDPE	Chemically inert, flexible, insulator	Squeeze bottles, toys, flexible pipes, insulation cover (electric wires), six pack rings, etc.
High-density polyethylene	HDPE	Inert, thermally stable, tough and high tensile strength[50]	Bottles, pipes, inner insulation (dielectric) of coax cable, plastic bags, etc.

Polymer	Abbreviation	Properties	Uses
Polypropylene	PP	Resistant to acids and alkalies, High tensile strength	Auto parts, industrial fibers, food containers, liner in bags, dishware and as a wrapping material for textiles and food
Polystyrene (thermocole)	PS	Thermal insulator. Properties depends on the form, expanded form is tough and rigid	Petri dishes, CD case, plastic cutlery
Polytetrafluoroethylene	PTFE	Very low coefficient of friction, excellent dielectric properties, chemically inert	Low friction bearings, non-stick pans, inner insulation (dielectric) of coax cable, coating against chemical attack etc.
Polyvinyl chloride	PVC	Insulator, flame retardant, chemically inert	Pipe (mainly draining), fencing, lawn chairs, hand-bags, curtain clothes, non-food bottles, raincoats, toys, vinyl flooring, electrical installation insulations, etc.

Polymers

Polymer	Abbreviation	Properties	Uses
Polychlorotrifluoroethylene	PCTFE	Stable to heat and thermal attacks, high tensile strength and non wetting	valves, seals, gaskets etc.

Brand names

These polymers are often better known through their brand names, for instance:

Brand Name	Polymer	Characteristic properties	Uses
Bakelite	Phenol-formaldehyde resin	High electric, heat and chemical resistance	Insulation of wires, manufacturing sockets, electrical devices, brake pads, etc.
Kevlar	Para-aramid fibre	High tensile strength	Manufacturing armour, sports and musical equipment. Used in the field of cryogenics[51]
Twaron	Para-aramid	Heat resistant and strong fibre	Bullet-proof body armor, helmets, brake pads, ropes, cables and optical fibre cables, etc. and as an asbestos substitute

Brand Name	Polymer	Characteristic properties	Uses
Mylar	Polyethylene terephthalate film	High strength and stiffness, less permeable to gases, almost reflects light completely	Food packaging, transparent covering over paper, reflector for rollsigns and solar cooking stoves
Neoprene	Polychloroprene	Chemically inert	Manufacturing gaskets, corrosion resistant coatings, waterproofseat covers, substitute for corks
Nylon	Polyamide	Silky, thermoplastic and resistant to biological and chemical agents	Stockings, fabrics, toothbrushes. Molded nylon is used in making machine screws, gears etc.
Nomex	Meta-aramid polymer	Excellent thermal, chemical, and radiation resistance, rigid, durable and fireproof.	Hood of firefighter's mask, electrical lamination of circuit boardsand transformer cores and in Thermal Micrometeoroid Garment
Orlon	Polyacrylonitrile (PAN)	Wool-like, resistant to chemicals, oils, moths and sunlight	Used for making clothes and fabrics like sweaters, hats, yarns, rugs, etc., and as a precursor of carbon fibres

Brand Name	Polymer	Characteristic properties	Uses
Rilsan	Polyamide 11 & 12	Bioplastic	Used in high-performance applications such as sports shoes, electronic device components, automotive fuel lines, pneumatic airbrake tubing, oil and gas flexible pipes and control fluid umbilicals, and catheters.
Technora	Copolyamid	High tensile strength, resistance to corrosion, heat, chemicals and saltwater	Used for manufacturing optical fiber cables, umbilical cables, drumheads, automotive industry, ropes, wire ropes and cables
Teflon	Polytetrafluoroethylene(PTFE)	Very low coefficient of friction, excellent dielectricproperties, high melting, chemically inert	Plain bearings, gears, non-stick pans, etc. due to its low friction. Used as a tubing for highly corrosive chemicals.
Ultem	Polyimide	Heat, flame and solvent resistant. Has high dielectric strength	Used in medical and chemical instrumentation, also in guitar picks

Brand Name	Polymer	Characteristic properties	Uses
Vectran	aromatic polyester	High thermal and chemical stability. Golden color. Has high strength, low creep, and is moisture resistant	Used as reinforcing fibres for ropes, cables, sailcloth. Also used in manufacturing badminton strings, bike tires and in electronics applications. Is the key component of a line of inflatable spacecraft developed by Bigelow Aerospace
Viton	Polytetrafluoroethylene (PTFE)	Elastomer	Depends on the grade of the polymer. Viton B is used in chemical process plants and gaskets.
Zylon	poly-p-phenylene-2,6-benzobisoxazole(PBO)	Very high tensile strength and thermal stability	Used in tennis racquets, table tennis blades, body armor, etc.

CHAPTER 13

Warfare and Defence

Chemical Warfare

Chemical warfare which is often abbreviated simply as 'CW' refers to usage of properties of chemical substances which are toxic in nature. Chemical warfare consists of using these toxic properties as weapons. There have always been huge debates on reasons for its usage. Many countries insist that these are used as a mode of protection which is mostly not the case.

Chemical warfare is different from the use of conventional weapons or nuclear weapons because the destructive effects of chemical weapons are not primarily due to any explosive force. The offensive use of living organisms (such as anthrax) is considered *biological warfare* rather than chemical warfare; however, the use of nonliving toxic products produced by living organisms (e.g. toxins such as botulinum toxin, ricin, and saxitoxin) is considered *chemical warfare* under the provisions of the Chemical Weapons Convention.

About 70 different chemicals have been used or stockpiled as chemical warfare agents during the 20th century. The entire class known as Lethal Unitary Chemical Agents and Munitions have been scheduled for elimination by the CWC.

Under the Convention, chemicals that are toxic enough to be used as chemical weapons, or that may be used to manufacture such chemicals, are divided into three groups according to their purpose and treatment:

- ❖ Schedule 1 – Have few, if any, legitimate uses. These may only be produced or used for research, medical, pharmaceutical or protective purposes (i.e. testing of chemical weapons sensors and protective clothing). Examples include nerve agents, ricin, lewisite and mustard gas. Any production over 100 g must be reported to the OPCW[52] and a country can have a stockpile of no more than one ton of these chemicals.

- ❖ Schedule 2 – Have no large-scale industrial uses, but may have legitimate small-scale uses. Examples include dimethyl methylphosphonate, a precursor to sarin also used as a flame retardant, and thiodiglycol, a precursor chemical used in the manufacture of mustard gas but also widely used as a solvent in inks.

- ❖ Schedule 3 – Have legitimate large-scale industrial uses. Examples include phosgene and chloropicrin. Both have been used as chemical weapons but phosgene is an important precursor in the manufacture of plastics and chloropicrin is used as a fumigant. The OPCW must be notified of, and may inspect, any plant producing more than 30 tons per year.

The use of toxic chemicals as weapons dates back thousands of years, but the first large scale use of chemical

Warfare and Defence

weapons was during World War I. They were primarily used to demoralize, injure, and kill ntrenched defenders, against whom the indiscriminate and generally very slow-moving or static nature of gas clouds would be most effective. The types of weapons employed ranged from disabling chemicals, such as tear gas, to lethal agents like phosgene, chlorine, and mustard gas.

Nuclear Weapons

A nuclear weapon (also called an atom bomb, nuke, atomic bomb, nuclear warhead, A-bomb, or nuclear bomb) is an explosive device that derives its destructive force from nuclear reactions, either fission (fission bomb) or from a combination of fission and fusion reactions (thermonuclear bomb). Both bomb types release large quantities of energy from relatively small amounts of matter.

The first test of a fission ("atomic") bomb released an amount of energy approximately equal to 20,000 tons of TNT[53] (84 TJ). The first thermonuclear ("hydrogen") bomb test released energy approximately equal to 10 million tons of TNT.

Nuclear Weapon

A thermonuclear weapon weighing little more than 2,400 pounds (1,100 kg) can release energy equal to more than 1.2 million tons of TNT . A nuclear device no larger than traditional bombs can devastate an entire city by blast, fire, and radiation. Since they are weapons of mass destruction, the proliferatio of nuclear weapons is a focus of international relations policy.

Warfare and Defence || 123 ||

Nuclear weapons have been used twice in war, both times by the United States against Japan near the end of World War II. On August 6, 1945, the U.S. Army Air Forces detonated a uranium gun-type fission bomb nicknamed "Little Boy" over the Japanese city of Hiroshima; three days later, on August 9, the U.S. Army Air Forces detonated a plutonium implosion-type fission bomb nicknamed "Fat Man" over the Japanese city of Nagasaki. These bombings caused injuries that resulted in the deaths of approximately 200,000 civilians and military personnel.

The mushroom cloud of the atomic bombing of the Japanese city of Nagasaki on August 9, 1945, rose some 11 mi (18 km) above the bomb›s hypocenter.

CHAPTER 14

Fats and Oils

Fats are one of the three main macronutrients, along with carbohydrates and proteins. Fat molecules consist of primarily carbon and hydrogen atoms and are therefore hydrophobic and are soluble in organic solvents and insoluble in water. Examples include cholesterol, phospholipids, and triglycerides.

The terms *lipid, oil,* and *fat* are often confused. *Lipid* is the general term, though a lipid is not necessarily a triglyceride. *Oil* normally refers to a lipid with short or unsaturated fatty acid chains that is liquid at room temperature, while *fat* specifically refers to lipids that are solids at room temperature.

Fat is an important foodstuff for many forms of life, and fats serve both structural and metabolic functions. They are a necessary part of the diet of most heterotrophs[54] (including humans) and are the most energy dense, thus the most efficient form of energy storage.

Some fatty acids that are set free by the digestion of fats are called essential because they cannot be synthesized in the body from simpler constituents. There are two essential fatty acids (EFAs) in human nutrition: alpha-linolenic acid (an omega-3 fatty acid) and linoleic acid (an omega-6 fatty acid). Other lipids needed by the body can be synthesized from these and other fats. Fats and other lipids are broken down in the body by enzymes called lipases produced in the pancreas. Fats are broken down in the healthy body to release their constituents, glycerol and fatty acids. Glycerol itself can be converted to glucose by the liver and so become a source of energy.

Bottles containing different vegetable oils

Butter containing saturated fat

(Example of unsaturated fat)

Types of fats in food

- **Saturated fat:-** A saturated fat is a type of fat in which the fatty acid chains have all or predominantly single bonds.

  Examples of foods containing a high proportion of saturated fat include animal fat products such as cream, cheese, butter, other whole milk dairy products and fatty meats which also contain dietary cholesterol. Certain vegetable products have high saturated fat content, such as coconut oil and palm kernel oil. Many prepared foods are high in saturated fat content, such as pizza, dairy desserts etc.

- **Unsaturated fat:-** An unsaturated fat is a fat or fatty acid in which there is at least one double bond within the fatty acid chain. A fatty acid chain is monounsaturated if it contains one double bond, and polyunsaturated if it contains more than one double bond. In biochemistry and nutrition, monounsaturated fatty acids (abbreviated MUFAs, or more plainly monounsaturated fats) are fatty acids that have one double bond in the fatty acid chain with all of the remainder carbon atoms being single-bonded. By contrast, polyunsaturated fatty acids (PUFAs) have more than one double bond.

- **Monounsaturated fat:-** Monounsaturated fats are found in animal flesh such as red meat, whole milk products, nuts, and high fat fruits such as olives and avocados. Olive oil is about 75% monounsaturated fat. The high oleic variety sunflower oil contains as least 70% monounsaturated fat. Canola oil and cashews are both about 58% monounsaturated fat.Other sources include avocado oil, macadamia nut oil, grapeseed oil, groundnut oil (peanut oil), sesame oil, corn oil, popcorn, whole grain wheat, cereal, oatmeal, almond oil, sunflower oil, hemp oil, and tea-oil .

- ❖ ***Polyunsaturated fat:-*** Polyunsaturated fat can be found mostly in nuts, seeds, fish, seed oils, and oysters etc

- ❖ ***Trans fat:-*** Trans fat, also called trans-unsaturated fatty acids or trans fatty acids, is a type of unsaturated fat that occurs in small amounts in meat and milk fat. It became widely produced industrially from vegetable and fish oils in the early 20th century for use in margarine and later also in snack food, packaged baked goods, and for frying fast food. Trans fats also occur naturally, e.g., the vaccenic acid in breast milk, and some isomers of conjugated linoleic acid (CLA). These trans fats occur naturally in meat and dairy products from ruminants. Butter, for example, contains about 3% trans fat.

- ❖ ***Interesterified fat:-*** Interesterified fat is a type of oil where the fatty acids have been moved from one triglyceride molecule to another. This is generally done to modify the melting point, slow rancidification[55] and create an oil more suitable for deep frying or making margarine[56] with good taste and low saturated fat content.

CHAPTER 15

New High Performance Materials

Carbon Fibres

Carbon fibers or carbon fibres (alternatively CF, graphite fiber or graphite fibre) are fibers about 5–10 micrometres in diameter and composed mostly of carbon atoms. Carbon fibers have several advantages including high stiffness, high tensile strength, low weight, high chemical resistance, high temperature tolerance and low thermal expansion. These properties have made carbon fiber very popular in aerospace, civil engineering, military, and motorsports, along with other competition sports. However, they are relatively expensive when compared with similar fibers, such as glass fibers or plastic fibers.

Carbon fibers are usually combined with other materials to form a composite. When impregnated with a plastic resin and baked it forms carbon-fiber-reinforced polymer (often referred to as carbon fiber) which has a very high strength-to-weight ratio, and is extremely rigid although somewhat brittle. Carbon fibers are also composited with other materials, such as graphite, to form reinforced carbon-carbon composites, which have a very high heat tolerance.

Tail of an RC helicopter, made of carbon fiber reinforced polymer

Motorcycle racing gloves with carbon fiber protectors for ligaments in fingers.

Ceramics

Scientifically, ceramics are inorganic and non-metallic solids. These materials are prepared by heating and subsequent cooling of basic inorganic and non-metallic materials. The heating and cooling action imparts the desired mechanical properties to the ceramics. A series of permutations and combinations are possible for heating temperatures and cooling timings. These combinations impart varying levels of hardness, ductility[57], strength, malleability[58], etc. to the ceramics.

The science of ceramics has evolved and established its significance to the modern engineering industry. Due to the ability to have high melting points, low conductivity of thermal and electrical charges, resistance to chemical reactions, etc. Ceramics are preferred and developed for complex applications. Their light weight and ease of use make them an ideal choice for both high end and low end Industrial applications.

Ceramics have transcended their usefulness to human race from earthen pots to modern aerospace applications. From the field of electronics to avionics and mechanical to electrical conduits, ceramics are widely used and preferred in Industrial applications. The semiconductor industry for example, benefits greatly from the low electrical conductivity of ceramics. Modern ICs[59] are expected to be faster and smaller at the same time. This puts immense pressure on the material and Industrial ceramics stand the test of time for such unique applications. Ceramic manufacturing standards are established by the industry to ensure that desired outputs are modelled as per standard requirements.

Engineering ceramics are preferred for mass and batch production in industries. The non-corrosion and non-reaction ability of ceramics make them an ideal candidate for castings. The liquid metals can be poured in the castings made of ceramics without the vessels reacting with the poured element. Also, due to low thermal conductivity coefficient, the ceramic containers are easier to handle.

Due to their light weight, the ceramics are finding huge applications in aviation industry. Also, due to the smooth surface finishes that they can bear without hampering the inherence qualities expected from the material, Automobile industry uses Industrial ceramics to great extent. In manufacturing industry, ceramic tiles are used as fire brick lining materials in boilers. The tiles are designed and prepared in such a way that they can withstand temperatures in excess of 1800oC. Due to the non-thermal conductivity, the ceramic materials become an obvious choice for furnaces.

Microalloys

Microalloyed steel is a type of alloy steel that contains small amounts of alloying elements (0.05 to 0.15%), including niobium, vanadium, titanium, molybdenum, zirconium, boron, and rare-earth metals[60]. Microalloyed steels are designed to provide better mechanical properties and/or greater resistance to atmospheric corrosion than conventional carbon steels. They are not considered to be alloy steels in the normal sense because they are designed to meet specific mechanical properties rather than a chemical composition. Microalloyed steels have been developed originally for large diameter oil and gas pipelines.

In terms of performance and cost, microalloyed steels are between a carbon steel and a low alloy steel. Their yield strength is between 275 and 750 MPa without heat treatment. Weldability is good, and can even be improved by reducing carbon content while maintaining strength. They provide lower fabrication costs over hot-rolled carbon steel from weight savings, and their reduced carbon content improves weldability and weldment toughness. When it comes to the oil and gas industry, they bring increased pumping capacity to in-line pipe, which leads to operational savings. Fatigue life[61] and wear resistance are superior to similar heat-treated steels. The disadvantages are that toughness are not as good as quenched and tempered (Q&T) steels. They must also be heated hot enough for all of the alloys to be in solution; after forming, the material must be quickly cooled to 540 to 600 °C.

CHAPTER 16

Fuels

A fuel is any material that can be made to react with other substances so that it releases energy as heat energy or to be used for work. The concept was originally applied solely to those materials capable of releasing chemical energy but has since also been applied to other sources of heat energy such as nuclear energy (via nuclear fission and nuclear fusion).

The heat energy released by reactions of fuels is converted into mechanical energy via a heat engine. Other times the heat itself is valued for warmth, cooking, or industrial processes, as well as the illumination that comes with combustion. Fuels are also used in the cells of organisms in a process known as cellular respiration, where organic molecules are oxidized to release usable energy.

Chemical

Chemical fuels are substances that release energy by reacting with substances around them, most notably by the process of combustion.

General types of chemical fuels

	Primary (natural)	Secondary (artificial)
Solid fuels	wood, coal, peat[62], dung, etc.	coke, charcoal
Liquid fuels	petroleum	diesel, gasoline, kerosene, LPG, coal tar, naphtha, ethanol
Gaseous fuels	natural gas	hydrogen, propane, methane, coal gas, water gas, blast furnace gas, coke oven gas, CNG

A Gasoline or Petrol Station

Biofuel

Biofuel can be broadly defined as solid, liquid, or gas fuel consisting of, or derived from biomass. Biomass can also be used directly for heating or power—known as biomass fuel. Biofuel can be produced from any carbon source that can be replenished rapidly e.g. plants. Many different plants and plant-derived materials are used for biofuel manufacture.

Recently biofuels have been developed for use in automotive transport (for example Bioethanol and Biodiesel), but there is widespread public debate about how carbon efficient these fuels are.

Fossil Fuels

Extraction of petroleum⬛Fossil fuels are hydrocarbons, primarily coal and petroleum (liquid petroleum or natural gas), formed from the fossilized remains of ancient plants and animals by exposure to high heat and pressure in the absence of oxygen in the Earth's crust over hundreds of millions of years.

Fossil fuels contain high percentages of carbon and include coal, petroleum, and natural gas.

Nuclear Fuels

Nuclear fuel is any material that is consumed to derive nuclear energy. Technically speaking, all matter can be a nuclear fuel because any element under the right conditions will release nuclear energy, but the materials commonly referred to as nuclear fuels are those that will produce energy without being placed under extreme duress. Nuclear fuel is a material that can be 'burned' by nuclear fission or fusion to derive nuclear energy. Nuclear fuels contain heavy fissile elements[63] that are capable of nuclear fission. When these fuels are struck by neutrons, they are in turn capable of emitting neutrons when they break apart. This makes possible a self-sustaining chain reaction that releases energy with a controlled rate in a nuclear reactor or with a very rapid uncontrolled rate in a nuclear weapon.

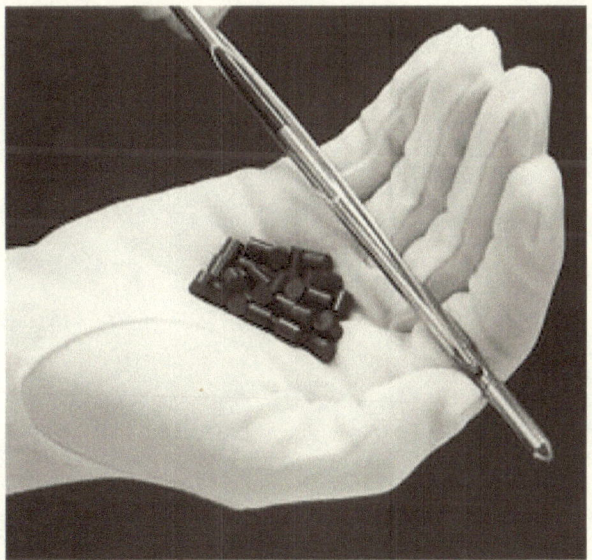

(Nuclear fuel pellets are used to release nuclear energy.)

The most common fissile nuclear fuels are uranium-235 (^{235}U) and plutonium-239 (^{239}Pu).

Fuels that produce energy by the process of nuclear fusion are currently not utilized by humans but are the main source of fuel for stars.

Liquid fuels for transportation

Most transportation fuels are liquids, because vehicles usually require high energy density. This occurs naturally in liquids and solids. High energy density can also be provided by an internal combustion engine. These engines require clean-burning fuels. The fuels that are easiest to burn cleanly are typically liquids and gases. Thus, liquids meet the requirements of being both energy-dense and clean-burning. In addition, liquids (and gases) can be pumped, which means handling is easily mechanized, and thus less laborious.

CHAPTER 17

Chemistry at Crime Scene/ Forensic Chemistry

Inspection at crime scene

Forensic chemistry is the application of chemistry and its subfield, forensic toxicology, in a legal setting. A forensic chemist can assist in the identification of unknown materials found at a crime scene. Specialists in this field have a wide array of methods and instruments to help identify unknown substances. These include high-performance liquid chromatography[64], gas chromatography-mass spectrometry, atomic absorption spectroscopy[65], Fourier transform infrared spectroscopy[66], and thin layer chromatography[67]. The range of different methods is important due to the destructive nature of some instruments and the number of possible unknown substances that can be found at a scene. Forensic chemists prefer using nondestructive methods first, to preserve evidence and to determine which destructive methods will produce the best results.

Chemicals used to reveal fingerprints

In many crime labs, there are four kinds of chemical reagents used to expose invisible, or latent, fingerprints. They are cyanoacrylate, silver nitrate, iodine, and ninhydrin.

What is "cyanoacrylate"?

Perhaps we know this item by its trade name, Super Glue. It is this same Super Glue that we can purchase at any Home Depot. Cyanoacrylate is the active ingredient that makes up 98% of Super Glue. This item has become a very practical and innovative forensic tool. When cyanoacrylate is heated or mixed with sodium hydroxide (NaOH), it releases fumes that interact with the amino acids that are found in the fingerprint residues found on an object, thus making a white print.

After exposure to cyanoacrylate, the fingerprints can then be captured on film as it is or treated with a fluorescent pigment that sticks to the fingerprint. The fingerprint then fluoresces, or glows, under a laser or ultraviolet light source.

An object that is suspected to have latent fingerprints in this method many times is exposed to the fumes inside a gadget called a fuming chamber. The end result is that the fumed fingerprints are now hard and stable as one would expect from Super Glue. In place of setting up a fuming chamber at the scene of a crime, CSI[68] technicians often times use a hand held wand-shaped tool that heats up a small cartridge of cyanoacrylate mixed together with fluorescent pigment. This tool then releases gases in close proximity of the latent prints, allowing the technician to fix and dye the fingerprint simultaneously.

"*Silver nitrate*" is a chemical ingredient found in black-and-white photographic film. When CSI technicians apply silver nitrate to a latent fingerprint, the chloride found in fingerprint residue interacts with the silver nitrate and forms another compound called silver chloride. This new compound reveals a black or reddish-brown fingerprint in the presence of ultraviolet light.

The third chemical used to reveal latent fingerprints is called "*iodine*". When heated up, crystalline iodine releases iodine fumes into a fuming chamber, where the iodine interacts with the oils found in the latent print thus producing a brownish colored fingerprint. Unfortunately, this kind of print has a tendency to fade rather quickly. Therefore, it must be captured on film right away or fixed by spraying it with a "fixing solution" made of water and starch. This fixing solution allows the print to last for weeks or even months in lieu of a few minutes.

The fourth kind of chemical reagent used to reveal latent fingerprints is "*ninhydrin*". Ninhydrin, also known as triketohydrindene hydrate, has been used for many years by CSI technicians to reveal latent fingerprints. An object suspected of containing latent fingerprints is sprayed with a solution of ninhydrin. It may take several hours for the fingerprints to show up due to the fact that ninhydrin reacts very slowly with the oils found in the fingerprint. However, heating up the object to a temperature of 80 to 100 degrees Fahrenheit can reduce the reaction time. The resulting fingerprint will be a purple/blue print.

Now, we have been educated on the kinds of chemicals used on our favorite *CSI* show. The next time we hear any of these chemical tools used on TV, we will know exactly what they are talking about.

Taking a Closer Look at the Evidence of a Crime Scene

After CSI (crime scene investigation) technicians have recovered suspicious material evidence like a possible accelerant at a crime scene suspected as arson[69], they bring it back to the crime lab and look at the evidence more closely. The scientific procedures used in an actual crime lab are similar to those we see on the CSI TV shows. In this article, we will discuss some of the procedures used when CSI technicians examine evidence in closer detail.

In a scientific crime lab, the following procedures are used to analyze compounds up close:

- Gas chromatography
- Mass spectroscopy, and
- Infrared spectrophotometry

No doubt, we all have heard of these terminologies when watching our favorite *CSI* program.

"Gas chromatography" (GC) is a method used to identify what a compound is and how far each part of it moves through an inert gas, or noble gas. GC very quickly separates mixtures of compounds into individual parts in this way.

The way GC is performed is that a lab technician injects a liquid sample into one end of the column that is heated and vaporized. The vapor enters the column and flows with the moving inert gas until it reaches another gadget called a detector. Different compounds move at different speeds and thus separate from each other. During this change, the individual compounds arrive at the detector at different times. It is a scientific marathon for individual compounds. The detector then signals another gadget called a recorder that prints a graph, called a chromatograph, that denotes each of the compounds detected. Many times, the result of the GC are needed by technicians to determine the composition of the unknown substance. If they need more information, technicians can combine the results of the GC with the results of either infrared spectrophotometry or mass spectroscopy.

"Mass spectroscopy" (MS) is another technique used in a crime lab to determine a substance's chemical makeup. Since no two substances have the exact same chemical makeup, or chemical fingerprint, this gadget can be used to determine the makeup of one substance and compare it to another.

The mass spectroscope is a machine that showers a compound with a beam of high-energy electrons that cause the separation of that compound. The resulting individual fragments then pass through an electric or magnetic field that separates them according to their individual masses. The pattern of fragmentation of the unknown compound is then compared via computer to known patterns of fragmentation on record and scrutinized by a technician with the hope of determining a consistency that unveils the composition of the unknown compound.

The GC and MS can be used together. This conglomerate is called a "gas chromatograph-mass spectrometer". The GC feeds each separated gaseous substance into the MS which in turn determines the substance's mass spectrum. Just about any substance can be identified using this combination.

Finally, "infrared spectrophotometry" (IR) is a technique used to identify the absorption spectrum of a compound since each compound has a unique infrared absorption spectrum. This method determines the amount of infrared light absorbed by the compound of interest and results in a chemical fingerprint of that same compound.

IR can also be used in conjunction with GC, and the GC-IR combination very quickly gives results that are just as accurate as the ones given by the GC-MS combination.

Interpreting the Results

Even though technicians cannot always determine the exact manufacturer of a given compound, a comparative analysis between the unknown compound and those taken from suspected areas may help them trace the unknown sample to a particular gas station or hardware store that sells the accelerant.

Though this particular investigative technique is not totally reliable, the investigative criminal scientist may be able to say that it came from a particular source.

The next time we watch any of the *CSI* shows, we will have better knowledge of what some of these techniques are used for whenever we hear the terminologies we have discussed.

CHAPTER 18

Potable Water

Drinking water, also known as potable water, is water that is safe to drink or to use for food preparation.

Parameters for drinking water quality typically fall within three categories:

- physical
- chemical
- microbiological

Physical and chemical parameters include heavy metals, trace organic compounds, total suspended solids (TSS)[70], and turbidity[71].

Microbiological parameters include Coliform bacteria, *E. coli*, and specific pathogenic species of bacteria (such as cholera-causing *Vibrio cholerae*), viruses, and protozoan parasites.

Chemical parameters tend to pose more of a chronic health risk through buildup of heavy metals although some components like nitrates/nitrites and arsenic can have a more immediate impact.

Different processes to clean drinking water

Treatment for drinking water production involves the removal of contaminants from raw water to produce water that is pure enough for human consumption without any short term or long term risk of any adverse health effect.

Processes

Empty aeration tank for iron precipitation

A combination selected from the following processes is used for municipal drinking water treatment worldwide:

- Pre-chlorination for algae control and arresting biological growth
- Aeration along with pre-chlorination for removal of dissolved iron when present with small amounts relatively of manganese
- Coagulation for flocculation[72] or slow-sand filtration
- Coagulant aids, also known as polyelectrolytes – to improve coagulation and for more robust floc formation
- Sedimentation[73] for solids separation that is the removal of suspended solids trapped in the floc
- Filtration to remove particles from water either by passage through a sand bed that can be washed and reused or by passage through a purpose designed filter that may be washable.
- Disinfection for killing bacteria, viruses and other pathogens.

Chemical cleaning is a method relevant in freeing walls and surfaces of various equipment, heat exchangers, pipelines, vessels and kettles from unwanted residues and bacterial contaminants. It is also essential in water purification, treatment, disinfection and conditioning. With varying methods to its name, several industries benefit from the procedure.

The most important facets of chemical cleaning gear on providing environment-friendly alternatives to consumers. No matter where it is applied, the aim is to primarily free the surface from contaminants. Below are the different methods listed under this cleaning process.

Chlorine dioxide method

Biofilms and legionella are two of the most common microbes that reside in water. Biofilms are residences or havens for organisms that contaminate products in water solutions. Legionnaires disease is caused by legionella or a Gram negative bacterium.

With the aid of chlorine dioxide, the biofilm or legionella may be removed from the water system. It is a favored technique because it has the capacity to prevent future formation of contaminants plus it can also be used in both hot and cold water systems. Chlorine is proven to have a safer solubility in water and has the capability to destroy THM precursors[74]. It also increases coagulation while it destroys phenols. It is not hazardous to health and does not have any distinct smell.

Ultraviolet radiation

Another proven chemical cleaning process for water disinfection is the ultraviolet or UV radiation. Aside from water, it may also disinfect air and solid surfaces contaminated by microbes. Its capacity to disinfect was reported since biology and light waves were introduced in the fields of science research.

Ultraviolet radiation has several advantages. It is considered safe because there is no need to monitor any dangerous chemicals; thus the results of disinfection are quickly attained. Any household or business establishment does not need to spend a lot since the initial system and operating costs are very minimal. It does not change the taste and smell of water as well plus they are easily installed and maintained.

Ozone purification

Ozone is one molecule that contains three atoms of oxygen and is naturally produced by a series of chemical reactions. It may come from the ultraviolet rays of the sun as well as from waterfalls and thunderstorms. It is also a chemical cleaning process used in wide areas of water treatment and purification. It is used in potable water treatments and in the broader municipal wastewater sites. It is also being widely recognized in several industries.

Bromination and Iodinization

Bromine and iodine can also be used as disinfectants. However, chlorine in water is over three times more effective as a disinfectant against *Escherichia coli* than an equivalent concentration of bromine, and over six times more effective than an equivalent concentration of iodine. Iodine is commonly used for portable water purification, and bromine is common as a swimming pool disinfectant.

CHAPTER 19

Chemical Waste

Chemical waste is a waste that is made from harmful chemicals (mostly produced by large factories). Chemical waste may fall under regulations such as COSHH in the United Kingdom, or the Clean Water Act and Resource Conservation and Recovery Act in the United States. In the U.S., the Environmental Protection Agency (EPA) and the Occupational Safety and Health Administration (OSHA), as well as state and local regulations also regulate chemical use and disposal. Chemical waste may or may not be classed as hazardous waste. A **chemical hazardous** waste is a solid, liquid, or gaseous material that displays either a "Hazardous Characteristic" or is specifically "listed" by name as a hazardous waste. There are four characteristics; chemical wastes may have to be considered as hazardous. These are Ignitability[75], Corrosivity[76], Reactivity[77], and Toxicity[78]. This type of hazardous waste must be categorized as to its identity, constituents, and hazards so that it may be safely handled and managed.

Chemical Waste Disposal Guideline

Innocuous aqueous waste	Organic Solvent	Red List	Solid Waste
• Acid (pH<4)	• Chlorinated Example: DCM[79], Chloroform, Chlorobenzene etch.	• Compounds with transitional metals	• Lightly contaminated Example: Gloves, empty vials/centrifuge.
• Alkali (pH<10)	• Non-Cholronated Example: THF, ethyl acetate, hexane, toluene, methanol, etc.	• Biocides	• <u>Broken Glassware</u> Broken glassware are usually collected in plastic-lined cardboard boxes for landfilling. Due to contamination, they are usually not suitable for recycling.
• Harmless soluble inorganic salt		• Cyanides	
• Alcohol containing salt		• Mineral oils and hydrocarbons	
• Hypochlorite solution		• Poisonous organosilicon compounds	
• Fine (tlc grade) silica and alumina		• Metal phosphides	
		• Phosphorus element	
		• Fluorides and nitrites.	

Chemical Compatibility Guideline

Many chemicals may react adversely when combined. It is recommended that incompatible chemicals be stored in separate areas of the lab.

Acids should be separated from alkalis, metals, cyanides, sulfides, azides, phosphides, and oxidizers. The reason being, when combined acids with these type of compounds, violent exothermic reaction can occur possibly causing flammable gas, and in some cases explosions.

Oxidizers should be separated from acids, organic materials, metals, reducing agents, and ammonia. This is because when combined, oxidizers with these type of compounds become inflammable, and sometimes toxic compounds can occur.

Container compatibility

When disposing hazardous laboratory chemical waste, chemical compatibility must be considered. For safe disposal, the container must be chemically compatible with the material it will hold. Chemicals must not react with, weaken, or dissolve the container or lid. Acids or bases should not be stored in metal. Hydrofluoric acid should not store in glass. Gasoline (solvents) should not store or transport in lightweight polyethylene containers such as milk jugs. Moreover, the Chemical Compatibility Guidelines should be considered for more detailed information.

Requirements for chemical waste disposal

Packaging, labelling, storage are the three requirements for disposing chemical waste.

Packaging

For packaging, chemical liquid waste containers should only be filled up to 75% capacity to allow for vapour expansion and to reduce potential spills which could occur from moving overfilled containers. Container material must be compatible with the stored hazardous waste. Finally, wastes must not be packaged in containers that improperly identify other non existing hazards.

In addition to the general packaging requirements mentioned above, incompatible materials should never be mixed together in a single container. Solvent safety cans should be used to collect and temporarily store large volumes (10–20 litres) of flammable organic waste solvents, precipitates, solids or other non-fluid wastes should not be mixed into safety cans.

Labelling

Label all containers with the group name from the chemical waste category and an itemized list of the contents. All chemicals or anything contaminated with chemicals posing a significant hazard. All waste must be appropriately packaged.

Storage

When storing chemical wastes, the containers must be in good condition and should remain closed unless waste is being added. Hazardous waste must be stored safely prior to removal from the laboratory and should not be

Chemical Waste

allowed to accumulate. Container should be sturdy and leakproof, also has to be labelled. All liquid waste must be stored in leakproof containers with a screw- top or other secure lid. Snap caps, mis-sized caps, parafilm and other loose fitting lids are not acceptable. If necessary, waste material is transferred to a container that can be securely closed. Waste containers are kept closed except when adding waste. Secondary containment should be in place to capture spills and leaks from the primary container, segregate incompatible hazardous wastes, such as acids and bases.

CHAPTER 20

Chemistry inside Human Body

Composition of the Human Body

Body composition may be analyzed in various ways. This can be done in terms of the chemical elements present, or by molecular type e.g., water, protein, fats (or lipids), hydroxylapatite (in bones), carbohydrates (such as glycogen and glucose) and DNA. In terms of tissue type, the body may be analyzed into water, fat, connective tissue, muscle, bone, etc. In terms of cell type, the body contains hundreds of different types of cells, but notably, the largest *number* of cells contained in a human body (though not the largest mass of cells) are not human cells, but bacteria residing in the normal human gastrointestinal tract.

Chemistry inside Human Body

Elements

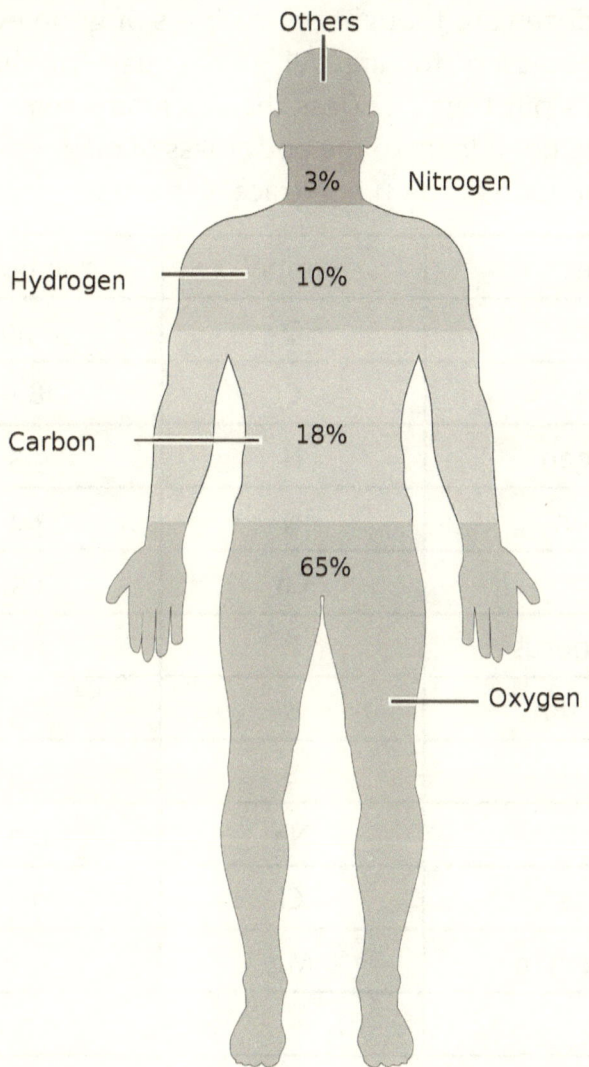

The main elements that compose the human body are shown from most abundant (by mass, not by fraction of atoms) to least abundant.

Almost 99% of the mass of the human body is made up of six elements: oxygen, carbon, hydrogen, nitrogen, calcium, and phosphorus. Only about 0.85% is composed of another five elements: potassium, sulfur, sodium, chlorine, and magnesium. All 11 are necessary for life. The remaining elements are trace elements, of which more than a dozen are thought on the basis of good evidence to be necessary for life. All of the mass of the trace elements put together (less than 10 grams for a human body) do not add up to the body mass of magnesium, the least common of the 11 non-trace elements.

Element	Symbol	% in body
Oxygen	O	65.0
Carbon	C	18.5
Hydrogen	H	9.5
Nitrogen	N	3.2
Calcium	Ca	1.5
Phosphorus	P	1.0
Potassium	K	0.4
Sulfur	S	0.3
Sodium	Na	0.2
Chlorine	Cl	0.2
Magnesium	Mg	0.2
Others		< 0.1

Molecules

The composition of the human body is expressed in terms of chemicals:

- ❖ Water
- ❖ Proteins – including those of hair, connective tissue, etc.

Chemistry inside Human Body

- Fats (or lipids)
- Hydroxyapatite in bones
- Carbohydrates such as glycogen and glucose
- DNA
- Dissolved inorganic ions such as sodium, potassium, chloride, bicarbonate, phosphate
- Gases such as oxygen, carbon dioxide, nitrogen oxide, hydrogen, carbon monoxide, acetaldehyde, formaldehyde, methanethiol. These may be dissolved or present in the gases in the lungs or intestines. Ethane and pentane are produced by oxygen free radicals.
- Many other small molecules, such as amino acids, fatty acids, nucleobases, nucleosides, nucleotides, vitamins, cofactors.
- Free radicals such as superoxide, hydroxyl, and hydroperoxyl.

Chemistry in Food Digestion

Digestion is the breakdown of large insoluble food molecules into small water-soluble food molecules so that they can be absorbed into the watery blood plasma. In certain organisms, these smaller substances are absorbed through the small intestine into the blood stream. Digestion is a form of catabolism that is often divided into two processes based on how food is broken down: mechanical and chemical digestion. The term **mechanical digestion** refers to the physical breakdown of large pieces of food into smaller pieces which can subsequently be accessed by digestive enzymes. In **chemical digestion**, enzymes break down food into the small molecules the body can use.

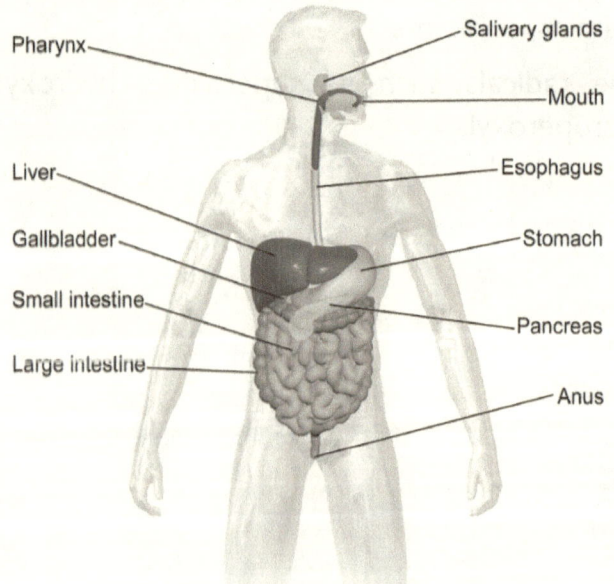

The Components of the Digestive System

In the human digestive system, food enters the mouth and mechanical digestion of the food starts by the action of mastication (chewing), a form of mechanical digestion, and the wetting contact of saliva. Saliva, a liquid secreted by the salivary glands, contains salivary amylase, an enzyme which starts the digestion of starch in the food; the saliva also contains mucus, which lubricates the food, and hydrogen carbonate, which provides the ideal conditions of pH (alkaline) for amylase to work. After undergoing mastication and starch digestion, the food will be in the form of a small, round slurry mass called a bolus. It will then travel down the esophagus and into the stomach by the action of peristalsis. Gastric juice in the stomach starts protein digestion. Gastric juice mainly contains hydrochloric acid and pepsin. In infants and toddlers gastric juice also contains rennin. As the first two chemicals may damage the stomach wall, mucus is secreted by the stomach, providing a slimy layer that acts as a shield against the damaging effects of the chemicals. At the same time protein digestion is occurring, mechanical mixing occurs by peristalsis, which is waves of muscular contractions that move along the stomach wall. This allows the mass of food to further mix with the digestive enzymes.

After some time (typically 1–2 hours in humans, 4–6 hours in dogs, 3–4 hours in house cats), the resulting thick liquid is called chyme. When the pyloric sphincter valve opens, chyme enters the duodenum where it mixes with digestive enzymes from the pancreas and bile juice from the liver and then passes through the small intestine, in which digestion continues. When the chyme is fully digested, it is absorbed into the blood. 95% of nutrient absorption occurs in the small intestine. Water and minerals are reabsorbed back into the blood in the colon (large intestine) where the pH is slightly acidic about 5.6 ~

6.9. Some vitamins, such as biotin and vitamin K produced by bacteria in the colon are also absorbed into the blood In the colon. Waste material is eliminated from the rectum during defecation.

Significance of pH

Digestion is a complex process controlled by several factors. pH plays a crucial role in a normally functioning digestive tract. In the mouth, pharynx and esophagus, pH is typically about 6.8, very weakly acidic. Saliva controls pH in this region of the digestive tract. Salivary amylase is contained in saliva and starts the breakdown of carbohydrates into monosaccharides. Most digestive enzymes are sensitive to pH and will denature in a high or low pH environment.

The stomach's high acidity inhibits the breakdown of carbohydrates within it. This acidity confers two benefits: it denatures proteins for further digestion in the small intestines, and provides non-specific immunity, damaging or eliminating various pathogens.

In the small intestines, the duodenum provides critical pH balancing to activate digestive enzymes. The liver secretes bile into the duodenum to neutralize the acidic conditions from the stomach, and the pancreatic duct empties into the duodenum, adding bicarbonate to neutralize the acidic chyme, thus creating a neutral environment. The mucosal tissue of the small intestines is alkaline with a pH of about 8.5

CHAPTER 21

Chemistry in Cooking

What is Cooking?

The term describes any process used to prepare food for eating. These can range quite widely. Methods such as ceviche[80] rely solely on plant acids to cook proteins. Pickling uses bacterial fermentation to preserve and prepare foodstuffs.

Pickling would serve to keep certain items (vegetables, for example) fresh much longer than otherwise possible, so people could stock up on food to last them during trying times, such as during winter.

However, cooking usually involves the application of heat in one way or another to prepare food. For the sake of simplicity, we will be talking about this last kind of cooking.

Carbohydrates

Chemically speaking, the carbs are molecules containing carbon, oxygen, and hydrogen, typically with oxygen to hydrogen ratio of 2:1. This is the same ratio as water. They're basically sugars and starches — which are also sugars, but with longer molecules. Sucrose (table sugar) is one such carb.

Chemistry in Cooking

If you've ever heated table sugar, you've noticed it will begin to brown, then liquify, and finally bubble. The bubbling occurs due to hydrogen and oxygen breaking off from the sugars, forming water molecules, and evaporating. The browning is owed to the polymerization of caramelans ($C_{24}H_{36}O_{18}$), caramelens ($C_{36}H_{50}O_{25}$), and caramelins ($C_{125}H_{188}O_{80}$). Caramel's unique aroma is given off by volatile substances released during their pyrolysis[81], such as diacetyl.

Why is caramelization important? Well, the process is key in cooking many plant-based foodstuffs. Vegetables cooked on high heat, in a stir-fry for example, will progressively brown as their starches and sugars break down and caramelize. Thankfully for us all, the volatiles released also give these veggies newfound 'yum'. Caramelization is also partly responsible for the golden-brown bakers' goods take on in the oven.

The Maillard reaction, a chemical reaction between amino acids and reducing sugars, is the other reaction that plays a role in baked goods' aroma, flavor, and golden hue. Like caramelization, it's also a type of heat-powered browning. However, it takes place at lower temperatures (at about 140 to 165 °C / 280 to 330 °F) and creates a wide-ranging cocktail of substances.

Lipids

Colloquially known as fats, they are long-chain hydrocarbons. Their exact chemical nuances are a bit complex, but suffice to say that the overwhelming majority of dietary fats are triglycerides. The longer these chains get, the more they will tend to jumble up. Because of this, at room temperature fats generally tend to be solids. Obviously, there are exceptions to this rule. Oils, for example, are made up of short-chained, mostly unsaturated fats. Apply heat, however, and fat melts — heat is energy, and enough energy allows fat's molecular chains to shake away from and start sliding past eachother, i.e. to flow.

Fats are one of the densest energy stores organisms produce. When heated to a liquid, fats are more easily absorbed by items of food, imbuing them with flavor. That's why fried food tastes awesome, why bakers put butter in their wares, and why cake is absolutely amazing (that and sugar of course).

Proteins

Protein denaturation is the real meat of chemical reactions in cooking. Proteins are incredibly complicated bits of molecular machinery. They're what imparts structural resilience to foodstuffs, what makes them chewy.

Proteins have four layers of complexity to them. First is the amino-acid composition itself, which forms the primary structure. Then come the intramolecular bonds between the amino-acids, the secondary structure, the shape the protein is folded in, the tertiary structure, and the overall macromolecule's 3D shape, the quaternary structure.

Cooking, however, breaks these layers of complexity down in a process we call denaturation. Essentially, this process reverts proteins to their primary, or at most secondary, structure. This makes the proteins much easier to break down chemically, and less able to hold together mechanically, which is why cooked food is easier to chew or digest than raw food and has more effective calories.

Chemistry in Cooking

In a way, the process is similar to that of carbs, as proteins are essentially long chains of mostly carbon and hydrogen. Outwardly, however, the effect is kind of the opposite of that in fats. An egg's white, for example, largely consists of proteins and water. Raw, it's gooey, runny, slimy, but holds together pretty well. When cooked, the denatured proteins start interacting with each other and jumbling up, making the white more firm but easier to bite and chew on.

In effect, cooking breaks down the egg white's proteins and then polymerizes and binds together. The same process takes place with all proteins in the food you **cook**, be it an egg or a slab of steak. The Maillard reaction again makes an appearance here. The reactions between proteins and carbohydrates that brown baked goods or toast are the same ones that make meats and other protein-heavy foods brown when cooked.

Taste

What you perceive as 'taste' is your brain interpreting the chemical composition of food. Sugars, for example, are sweet and fat gives foods that irresistible, savory flavor.

Cooking alters each of these macronutrients, carbs, fats, and proteins both separately and as a whole. On one hand, you have proteins being denatured, on the other you have sugars being transformed, and finally, you have interactions going on between the two in Maillard reactions. Lipids will also seep into the whole thing along with the cocktail of compounds released by sugars and proteins, soaking everything in delicious goodness.

The chemical processes that occur in the pan are extremely complex, and are not completely understood for a simple reason: no two foodstuffs are alike. We have a general idea of the main going-one while cooking, but it's simply too large to understand fully. However, the above information has helped us to get a whole new appreciation of wonderful chemistry that comes together in a tasty dish.

Chemical Reactions involved in Baking a Cake

We may not think of chemistry when we're baking a cake, but it is definitely a chemically based process. Whatever type of food we bake, the recipe's basic ingredients are involved in several chemical reactions that tie diverse ingredients together to form the finished dish.

Gluten Formation

Most baking is based on the use of flour, the powder form of grains, nuts and beans. Wheat flour, the most commonly used type of flour in baking, is composed largely of starch and protein, with very high levels of a class of protein known collectively as gluten. When water is added to wheat flour, the gluten forms a heavy, pliable mass. This expands greatly under hot temperatures and sets with the desired airy texture.

Leavening Agents

Leavening agents such as baking soda, baking powder and yeast give baked dough its lightness. Baking soda reacts with acids in the dough to make carbon dioxide, which helps the dough to rise. Baking powder, which is baking soda with an additional acidic salt, releases carbon dioxide twice during the baking process, once when it hits water, and again when it reaches a certain temperature in the oven. Heat helps baking powder produce tiny bubbles of carbon dioxide, which make a cake light and fluffy. When yeast, a single-celled fungus that feeds on starch and sugars, is added to dough, it also releases carbon dioxide bubbles, giving the dough a light, delicate texture.

Browning Reactions

Sugar does much more than just sweeten a cake. When the baking temperature reaches 300 degrees Fahrenheit, sugar undergoes what is known as a Maillard reaction, a chemical reaction between amino acids, proteins and reducing sugars. The result is browning, which forms the crust of many baked goods, such as bread. The Maillard reaction and caramelization[82], but both work together to create appetizing golden-brown surfaces and an array of flavors. Sugar also provides a direct food source for yeast, enhancing the yeast's activity.

Emulsification and Binding

Eggs in a cake mixture may fulfill one or more of three functions. Beaten egg white is used, like baking powder, to give the dough a light, fluffy consistency. This is possible because egg white (albumen) contains lecithin, a protein that coats the air bubbles created during beating, which stops the cake from sinking during baking. Lecithin also acts as a binder to hold the cake together. When egg is used as a glaze, it also acts as a source of protein for the sugar's Maillard reaction.

Onion Chemistry

Why does chopping onions make us cry?

Onions produce the chemical irritant known as syn-propanethial-S-oxide. It stimulates the eyes' lachrymal glands so they release tears. Scientists used to blame the enzyme allinase for the instability of substances in a cut onion. Recent studies from Japan, however, proved that lachrymatory-factor synthase, (a previously undiscovered enzyme) is responsible for the same.

The process goes as follows:

- Lachrymatory-factor synthase is released into the air when we cut an onion.
- The synthase enzyme converts the amino acids sulfoxides of the onion into sulfenic acid.
- The unstable sulfenic acid rearranges itself into syn-propanethial-S-oxide.
- Syn-propanethial-S-oxide gets into the air and comes in contact with our eyes. The lachrymal glands become irritated and produce the tears!

CHAPTER 22

Chemistry in Recreation

CHEMISTRY AND SPORTS

When you walk through a sporting goods store, you probably don't think about the technology behind the products on display in front of you. If you did, one material in particular would stand out: flexible PVC. The abbreviation for poly vinyl chloride, flexible PVC is used in a wide range of consumer goods, in particular those in the sporting goods category.

Some of the physical properties of PVC that make it such a versatile product are:

- PVC has high chemical and mechanical stabilities.
- PVC possess fire retardant capabilities.
- PVC is durable.
- PVC resists oils and chemicals.
- PVC can be molded and shaped.

In the sporting goods realm, the durability and the moldability of the material are most desirable. More specifically, the flexible PVC that is used in sporting shoes requires that it be formed into a shape that fits the foot. As for sporting equipment, most items are exposed to a lot of physical stress. PVC is so durable, it can stand up to all of the impact it receives. Sports shoes, like those used for running and on the basketball court, also make extensive use of flexible PVC to provide shock absorbance, support and strength in demanding use environments. Flexible PVC is found everywhere throughout a sporting goods store ranging from football and rugby, to golf and sailing.

Sports are one of the activities who have benefited the most from the development of new materials in chemistry labs. We all have noticed that sports equipment, from mountain bikes to surfboards, is becoming lighter and stronger.

We have also benefited from the advantages of modern sportswear, like fabrics that allow the rapid evaporation of sweat, or comfortable trainers with soles that adapts to the ground and the list goes on: cycling helmets, tennis racquets, football leggings, boat sails, athletics tracks. And, of course, balls! Golf, tennis, bowling, volleyball, football and basketball balls.

Chemistry in Recreation

Following are the few chemicals used in stadium equipments, chairs, roofs, turfs and sports equipments:-

- Polyolefins – These are used in modern day stadium turfs.
- Polycarbonate – These materials are used for stadium roofing purpose
- PVC – All most all parts of the stadiums are made up of PVC material.
- **Polyurethanes:-** Polyurethanes, a kind of plastic play an important role as they are frequently found in running and other athletic shoes, making them more resilient. In addition, polyurethane is found in a wide variety of popular sporting equipment, such as soccer balls, binders on running tracks and judo mats. A number of styles of sports flooring and pour-in-place track surfaces use polyurethanes, as well. These equipment necessities alongside such items as surfboard, roller blades, bowling balls and spandex apparel are all made possible in part due to polyurethane innovations.
- **Polycarbonate:-** Polycarbonate, a strong, shatter-resistant plastic, can also be found in protective sports equipment. Polycarbonate is often used in riding and biking helmets. Polycarbonate sunglasses and protective visors, which provide optical clarity as well as shatter-resistance, are worn by runners and rowers, just to name a few. Polycarbonate lenses can also be found in swim goggles.
- **Cyanoacrylate:-** Cyanoacrylate is the generic name for a family of fast-acting adhesives with industrial, medical and household uses. Cyanoacrylate is used in archery to glue fletching to arrow shafts.

- **Vectran:-** Vectran is a manufactured fiber, spun from a liquid crystal polymer (LCP). Vectran fiber is used in manufacturing badminton strings such as Yonex BG-85 and BG-80.
- In manufacturing of racquets the chemicals used are Graphite, Aluminum,, Boron, Nylon etc.
- Golf heads are made up of metal types like Zinc, Die Cast Aluminums, 6061 Aluminum, 304 Stainless Steel, Carbon Steel, 431 Stainless Steel, 17-4 Stainless Steel, 6-4 Titanium, 15-5 Stainless Steel, Beta Titanium etc.
- Different parts of cricket helmet are made up of Titanium, Polycarbonate, ABS[83] Plastic. Polyester etc.
- Pole vaulting pole is made up of carbon-fiber to promote higher jump.
- Mouth guards are made of rubber or flexible polyvinyl, which is highly durable and easy to mold.
- Sports Uniforms are made of a blend of synthetic fibers including nylon, polyester, rayon, and acetate.
- Shin guard - Shin guards are made from fiberglass, polyurethane, or foam rubber.

CHEMISTRY IN SWIMMING POOLS

Because swimming pools sometimes accommodate a variety of different people at a time, it is important to maintain a specific swimming pool chemistry in order to combat against the spread of disease and illness through the water. Properly maintaining the balance of chemicals in the pool environment can be a difficult task, especially if a pool is getting used by several different swimmers on a regular basis.

Below are the factors to know the fundamentals of having a well balanced water chemistry.

Total Alkalinity

The reason why total alkalinity[84] should be properly maintained is that it can greatly affect the pH level in the water. Without proper alkalinity balance, the pH level may consistently fluctuate. We can increase water's alkalinity by using a sodium bicarbonate solution and we can decrease the level by adding muriatic acid or sodium bisulfate. The desired range for alkalinity is 80-120 ppm or parts per million.

pH Level

The pH level is another factor we must consider when maintaining swimming pool water chemistry. If the pH level is not well balanced, swimmers may experience common discomforts and chlorine becomes useless. The ability of chlorine to oxidize matter and kill microorganisms is directly affected by pH. As the pH raises this ability is adversely affected.

In addition, at a pH over 8.0 scaling (precipitation of mineral components) and cloudy water may result. As pH falls below 7.0 the acidic condition will cause irritation to the eyes and mucous membranes of swimmers. Low pH (acidic water) can also corrode metal parts of a pool system and damage the plaster finish.

pH should be kept between 7.2-7.8. Human tears have a pH of 7.4, making this an ideal point to set a pool. We can increase the pH level (make it more alkaline) by adding caustic soda or soda ash. Conversely, we can decrease the pH level (make it more acidic) by adding sodium bisulfate.

Chlorine Compounds

The moment we add chlorine solution in the water, its strength percentage becomes dependent to the pH level of the water. With 7.0 pH level, 75% of chlorine is active hypochlorous acid. With a 7.5 pH level the activity of chlorine drops to 48%. At the undesirable pH level of 8.0, chlorine activity is only 22%. The proper range of chlorine for swimming pool water chemistry is 1.0 to 3.0 ppm.

The most popular pool disinfectant is the element chlorine, in the form of a chemical compound such as calcium hypochlorite (a solid) or sodium hypochlorite (a liquid). When the compound is added to the water, the chlorine reacts with the water to form various chemicals, most notably hypochlorous acid. Hypochlorous acid kills bacteria and other pathogens by attacking the lipids in the cell walls and destroying the enzymes and structures inside the cell through an oxidation reaction. Alternative sanitizers, such as bromide, do basically the same thing with slightly different results.

Chlorine is typically prepared in liquid, powder or tablet form and it can be added to the water anywhere in the cycle. Pool experts generally recommend adding it just after the filtering process, using a chemical feeder. If it's added directly into the pool, using tablets in the skimmer boxes, for example, the chlorine tends to be too concentrated in those areas.

Chloramines

When the chlorine compounds react with contaminants in the water, such as with ammonia and organic nitrogen compounds, chloramines are eventually formed. Essentially, the chlorine becomes useless as a disinfectant. As well, they are often the culprit of most eye irritation and odor problems. Chloramines can be eliminated easily by simply adding 10 ppm of free available chlorine compounds per ppm of chloramines.

Cyanuric Acid Level

Cyanuric acid or CYA is a stabilizing compound that prevents chlorine from being destroyed by the sun. If the level of CYA is too high, chlorine will eventually lose its sanitizing effect. With the proper level, the chlorine can last 5-10 times longer. The ideal range is 30-50 ppm although it is okay if it is 80-100 ppm.

Amount of Dissolved Solids

Basically, this factor refers to all the solid materials dissolved in the pool water. As the water is being reused and chemicals are being added, loads of hazardous materials develop and cause adversarial effects on swimming pool water chemistry. If this problem occurs, one only resort is to drain and refill pool with fresh, clean water. The proper range of total dissolved solids should not be more than 1,500 ppm.

Temperature

Temperature may not necessarily affect swimming pool water chemistry. However, it is important to make any swimming experiences enjoyable and refreshing. For typical enjoyment, the ideal temperature range is around 75 to 85 degrees Fahrenheit.

Photography

Photographic processing or photographic development is the chemical means by which photographic film or paper is treated after photographic exposure to produce a negative or positive image. Photographic processing transforms the latent image into a visible image, makes this permanent and renders it insensitive to light.

All processes based upon the gelatin-silver process are similar, regardless of the film or paper›s manufacturer. Exceptional variations include instant films such as those made by Polaroid and thermally developed films.

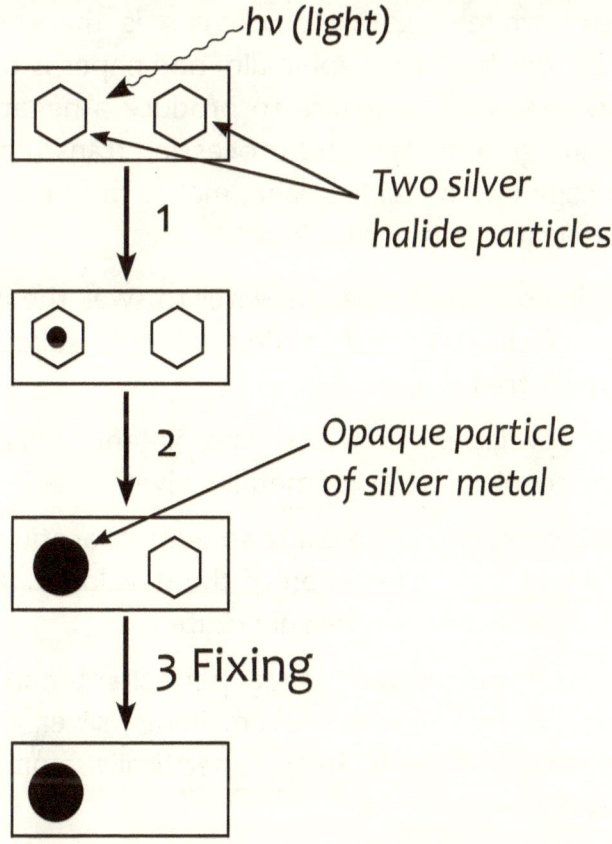

Key stages in production of Ag-based photographs

Two silver halide particles, one of which is impinged with light (hν) resulting in the formation of a latent image (step 1). The latent image is amplified using photographic developers, converting the silver halide crystal to an opaque particle of silver metal (step 2). Finally, the remaining silver halide is removed by fixing (step 3).

All photographic processing use a series of chemical baths. Processing, especially the development stages, requires very close control of temperature, agitation and time.

Black and white negative processing

Black and white negative processing is the chemical means by which photographic film and paper is treated after photographic exposure to produce a negative or positive image. Photographic processing transforms the latent image into a visible image, makes this permanent and renders it insensitive to light.

- ❖ The film may be soaked in water to swell the gelatin layer, facilitating the action of the subsequent chemical treatments.
- ❖ The developer converts the latent image to macroscopic particles of metallic silver.
- ❖ A stop bath, typically a dilute solution of acetic acid or citric acid, halts the action of the developer. A rinse with clean water may be substituted.
- ❖ The fixer makes the image permanent and light-resistant by dissolving remaining silver halide. A common fixer is hypo, specifically ammonium thiosulfate.

Chemistry in Recreation || 183 ||

- Washing in clean water removes any remaining fixer. Residual fixer can corrode the silver image, leading to discolouration, staining and fading.

The washing time can be reduced and the fixer more completely removed if a hypo clearing agent is used after the fixer.

- Film may be rinsed in a dilute solution of a non-ionic wetting agent to assist uniform drying, which eliminates drying marks caused by hard water. (In very hard water areas, a pre-rinse in distilled water may be required – otherwise the final rinse wetting agent can cause residual ionic calcium on the film to drop out of solution, causing spotting on the negative.)
- Film is then dried in a dust-free environment, cut and placed into protective sleeves.

Once the film is processed, it is then referred to as a *negative*.

The negative may now be printed; the negative is placed in an enlarger and projected onto a sheet of photographic paper. Alternatively (or as well), the negative may be scanned for digital printing or web viewing after adjustment, retouching, and/or manipulation.

Black and white reversal processing

This process has three additional stages:

- Following the stop bath, the film is bleached to remove the developed negative image. The film then contains a latent positive image formed from unexposed and undeveloped silver halide salts.
- The film is fogged, either chemically or by exposure to light.

- The remaining silver halide salts are developed in the second developer, converting them into a positive image.
- Finally, the film is fixed, washed, dried and cut.

Colour processing

Chromogenic materials use dye couplers to form colour images. Modern colour negative film is developed with the C-41 process and colour negative print materials with the RA-4 process. These processes are very similar, with differences in the first chemical developer.

The C-41 and RA-4 processes consist of the following steps:

- The colour developer develops the silver negative image, and by products activate the dye couplers to form the colour dyes in each emulsion layer.
- A rehalogenising bleach converts the developed silver image into silver halides.
- A fixer removes the silver salts.
- The film is washed, stabilised, dried and cut.

CHAPTER 23

Chemistry and Consumer Products

Chemistry in Cells/Batteries

Electricity is the flow of electrons through a conductive path called a circuit. Batteries consist of an anode (-), a cathode (+), and an electrolyte. The battery produces chemical reactions, causing a buildup of electrons at the anode, which results in an electrical difference between the anode and cathode. The electrons attempt to eliminate this difference by repelling each other and attempting to go somewhere with fewer electrons. In a battery, the only place for the electrons to go is to the cathode, but the electrolyte keeps them from doing this. When a wire connects the cathode to the anode, the circuit becomes closed, thus allowing the electrons into the cathode. This is how electrical potential causes electrons to flow through the circuit. In any battery, an electrochemical reaction moves electrons from one pole to the other. The metals and electrolytes in the battery control the voltage, with each different reaction causing a characteristic voltage. Modern batteries use a variety of chemistries including:

- **Alkaline** - commonly found in Energizer and Duracell brands, alkaline batteries have zinc and manganese-oxide electrodes and an alkaline electrolyte.
- **Lithium-iodide** - this type of battery chemistry is long lasting, and typically used in hearing aids and pacemakers.
- **Silver-zinc** - designed with a high power-to-weight ratio, silver-zinc batteries are most often used in aeronautical applications.
- **Zinc-carbon** (standard carbon battery) - is used in nearly all AA, C, and D dry-cell batteries. The electrodes are made entirely of zinc and carbon, with an acidic paste acting as the electrolyte.
- **Lead-acid** - used in automobile batteries, lead-acid chemistry contains lead and lead-oxide electrodes and a strong, rechargeable electrolyte.
- **Lithium-ion** - often found in high-end laptops and cell phones, this rechargeable battery chemistry offers excellent power-to-weight ratio.

Chemistry and Fireworks

We've all seen fireworks in lovely colors and effects but have you ever stopped to think how those wonderful colors are produced? To understand the chemistry of fireworks colors we will need knowledge of the structure of atoms and how electrons behave when they absorb energy. Let us learn about how the colours are produced when a firework is ignited.

The chemistry of fireworks colours

Let's start with the basic structure of the atom. Every atom has a nucleus that contains protons which are positively charged with negatively charged electrons in orbitals around the nucleus. The number of protons is equal to the number of electrons so that the atom overall is neutral. Now electrons will be in the lowest energy orbital[85] that is possible to be stable. This will be the orbital closest to the nucleus. Let's take lithium as an example. Lithium is the third element in the periodic table and has three protons and three electrons. Two electrons are in an orbital close to the nucleus. This orbital can only hold two electrons as it is has a small radius. So the third electron goes in an orbital a bit farther away from the nucleus. We say that lithium has an electronic configuration of 2,1.

Now there are lots more orbitals further away from the nucleus that are empty in the case of lithium. They are higher in energy as they are further from the pull of the nucleus so an electron would need more energy to stay in those orbitals. If we give that one electron some energy, for instance in the form of heat, then it can jump from its orbital into a higher orbital. Once there, it is unstable and falls back to its original starting point. However, when it falls back it has to lose that extra energy and it does this in the form of light. The light is emitted and the wavelength of the light depends on the difference between the energies of the starting and finishing orbitals. These will be different for different metals so the wavelength of the light will be different. For lithium the main transition that occurs; emits light that is red. Other metals emit light as shown below.

- Sodium — yellow/orange
- Potassium — lilac
- Calcium — brick red
- Barium — light green
- Copper — blues

So we can use metals in fireworks that, when they absorb energy from the burning charcoal, emit light of different colors. Other metals that can be added are magnesium which emits a brilliant white light, and aluminum and titanium which also produce white light and increase the temperature at which the firework burns.

So now, when we see a fireworks display, we will know how those colors are produced.

Shoe polish

Shoe polish is a waxy paste, cream, or liquid used to polish, shine, and waterproof leather shoes or boots to extend the footwear's life, and restore, maintain and improve their appearance.

Types

Shoe polish can be classified into three types: wax, cream-emulsion, and liquid. Each differs in detailed composition but all consist of a mixture of waxes, solvent, and dyes.

Wax-based shoe polish

Waxes, organic solvents and dyes compose this type of polish. Waxes are 20–40% of the material. Natural waxes include carnauba and montan as well as synthetic waxes. Solvents are selected to match the waxes. About 70% of shoe polish is solvent. A variety of solvents are used including naphtha. Turpentine, although more expensive, is favored for its "shoe polish odor". Dyes make up the final 2–3% of the polish. A traditional dye is nigrosine, but other dyes (including azo dyes) and pigments are used for oxblood, cordovan, and brown polishes..

Cream-Emulsion shoe polish

These polishes may have a gelatinous consistency. They are composed of the usual three components: waxes, liquid vehicle, and dyes. Unlike wax-based shoe polishes, cream-emulsions contain water and/or oil plus a solvent (either naphtha, turpentine or Stoddard Solution), so the liquid content is high.

Emulsifiers[86] and surfactants[87] are required. These include ammonia, morpholine and various ethoxylated surfactants such as polysorbate 80. The waxes are often some mixture of carnauba wax, beeswax, montan wax and its oxidized derivatives, and paraffin waxes.

Liquid shoe polish

Liquid shoe polish is sold in a squeezable plastic bottle, with a small sponge applicator at the end. To decrease its viscosity[88], bottled polish usually has a very low wax content. Liquid shoe polish is a complex mixture. Polyethylene wax emulsion is a major component.

Various polymers, typically acrylates, are the next major component, conferring gloss and holding the dyes in suspension. Resins and casein are selected to ensure adhesion to the leather. Fatty phosphate esters, emulsifiers, and glycols are also used. Pigments include *titanium dioxide* for whites and iron oxides for browns.

Chemistry and Sand

Brown sand dunes

Sand is a granular material composed of finely divided rock and mineral particles. It is defined by size, being finer than gravel[89] and coarser than silt[90]. Sand can also refer to a textural class of soil or soil type; i.e., a soil containing more than 85 percent sand-sized particles by mass.

The composition of sand varies, depending on the local rock sources and conditions, but the most common constituent of sand

is silica (silicon dioxide, or SiO_2), usually in the form of quartz. The second most common type of sand is calcium carbonate, for example, aragonite, which has mostly been created, over the past half billion years, by various forms of life, like coral and shellfish.

Sand is a non-renewable resource over human timescales, and sand suitable for making concrete is in high demand. Desert sand, although plentiful, is not suitable for concrete. 50 billion tons of beach sand and fossil sand is used each year for construction.

Chemistry in Safety Matches

A Lit Matchstick

There's a lot of interesting chemistry going on in the small head of a safety match. Safety matchsticks consist of a head and a wooden stick. The head usually contains *potassium chlorate*, an oxidizing agent, a small quantity of powdered glass that provides the essential friction while striking, animal glue to bind some other abrasives, and additive compounds like sulphur or antimony (III) sulfide that act as fuel.

The wooden stick also has a certain substance, typically *ammonium phosphate*, impregnated in its composition to suppress the afterglow once the flame dies. The other part is the striking surface, which usually consists of red phosphorus, powdered glass (or silica), binder and filler.

Working of Safety Matches

When we strike a safety match, the glass-on-glass friction generates heat, converting a small amount of red phosphorus to white phosphorus vapor.

White phosphorus spontaneously ignites, decomposing potassium chlorate and liberating oxygen.

At this point, the sulfur starts to burn, which ignites the wood of the match. The match head is coated with paraffin wax so the flame burns into the stick.

The wood of a match is special, too. Match sticks are soaked in an ammonium phosphate solution that reduces afterglow when the flame goes out.

Mosquito Coils

A mosquito coil is a mosquito-repelling incense, usually made into a spiral, and typically made using dried paste of pyrethrum powder.

Active ingredients found in mosquito coils may include:

- Pyrethrum – a natural, powdered material from a kind of chrysanthemum plant.
- Pyrethrins – an extract of the insecticidal chemicals in pyrethrum.
- Allethrin – sometimes d-trans-allethrin, the first synthetic pyrethroid.
- Esbiothrin – a form of allethrin.
- Meperfluthrin - a pyrethroid ester
- Butylated hydroxytoluene (BHT) – an optional additive used to prevent pyrethroid from oxidizing during burning.
- Piperonyl butoxide (PBO) – an optional additive to improve the effectiveness of pyrethroid.
- N-Octyl bicycloheptene dicarboximide (MGK 264) – an optional additive to improve the effectiveness of a pyrethroid.
- Dimefluthrin - a novel pyrethroid pesticide.
- Metofluthrin - a pyrethroid insecticide which is highly effective against mosquitoes. Also used in most mosquito repellants like candles, patches and mini fans.

Fountain Pen Inks

Fountain pen ink is a water-based ink intended for use with fountain pens. Fountain pen ink is almost exclusively dye-based.

Red inks usually contain the dye eosin. Blue inks often contain triarylmethane dye. Black inks are mixtures. In addition to water, the non-dye components (collectively referred to as the vehicle) might include polymeric resins, humectants to retard premature drying, pH modifiers, anti-foaming agents, biocides to prevent fungal and bacterial growth, and wetting agents (surfactants). Surfactants reduce the surface tension of the ink; distilled water has a surface tension of 72 dyn/cm (72 × 10^{-3} N/m), but the desirable value for ink is between 38 and 45 dyn/cm (38 to 45 × 10^{-3} N/m). If the ink's surface tension were too high, then it would not flow through the pen; if it were too low, then the ink would run out of the pen with less control.

Crayons

The term crayon is commonly associated with the standard wax crayon, such as those widely available for use by children. Such crayons are made mostly of paraffin wax. *Paraffin wax* (or petroleum wax) is a soft colorless solid, derived from petroleum, coal or shale oil, that consists of a mixture of hydrocarbon molecules containing between twenty and forty carbon atoms.

A crayon (or wax pastel) is a stick of colored wax, charcoal, chalk(Chalk is a soft, white, porous, sedimentary carbonate rock, a form of limestone composed of the mineral calcite. Calcite is an ionic salt called calcium carbonate or $CaCO_3$) or other material used for writing or drawing. A crayon made of pigment with a dry binder is a pastel; when made of oiled chalk, it is called an oil pastel.

CHAPTER 24

Chemistry and Atmosphere of Earth

The **atmosphere of Earth** is the layer of gases, commonly known as **air**, that surrounds the planet Earth and is retained by Earth's gravity. The atmosphere of Earth protects life on Earth by creating pressure allowing for liquid water to exist on the Earth›s surface, absorbing ultraviolet solar radiation, warming the surface through heat retention (greenhouse effect), and reducing temperature extremes between day and night (the diurnal temperature variation).

Composition

The three major constituents of Earth's atmosphere are nitrogen, oxygen, and argon. Water vapor accounts for roughly 0.25% of the atmosphere by mass. The concentration of water vapor (a greenhouse gas) varies significantly from around 10 ppm by volume in the coldest portions of the atmosphere to as much as 5% by volume in hot, humid air masses, and concentrations of other atmospheric gases are typically quoted in terms of dry air (without water vapor). The remaining gases are often referred to as trace gases, among which are the greenhouse gases, principally carbon dioxide, methane, nitrous oxide, and ozone. Besides argon, already mentioned, other noble gases, neon, helium, krypton, and xenon are also present. Filtered air includes trace amounts of many other chemical compounds. Many substances of natural origin may be present in locally and seasonally variable small amounts as aerosols[91] in an unfiltered air sample, including dust of mineral and organic composition, pollen and spores, sea spray, and volcanic ash. Various industrial pollutants also may be present as gases or aerosols, such as chlorine (elemental or in compounds), fluorine compounds and elemental mercury vapor. Sulfur compounds such as hydrogen sulfide and sulfur dioxide (SO_2) may be derived from natural sources or from industrial air pollution.

Major constituents of dry air, by volume

Gas		Volume	
Name	Formula	in ppmv (parts per million by volume)	in %
Nitrogen	N_2	780,840	78.084
Oxygen	O_2	209,460	20.946
Argon	Ar	9,340	0.9340
Carbon dioxide	CO_2	413.32	0.041332
Neon	Ne	18.18	0.001818
Helium	He	5.24	0.000524
Methane	CH_4	1.87	0.000187
Krypton	Kr	1.14	0.000114

Chemical Pollutants in the Environment

An air pollutant is a material in the air that can have adverse effects on humans and the ecosystem. The substance can be solid particles, liquid droplets, or gases. A pollutant can be of natural origin or man-made.

Anthropogenic (man-made) sources

These are mostly related to the burning of multiple types of fuel.

Stationary sources include smoke stacks of fossil fuel power stations (for example environmental impact of the coal industry), manufacturing facilities (factories) and waste incinerators, as well as furnaces and other types of fuel-burning heating devices. In developing and poor countries, traditional biomass burning is the major source of air pollutants; traditional biomass includes wood, crop waste and dung.

Mobile sources include motor vehicles, marine vessels, and aircraft.

Controlled burn practices in agriculture and forest management. Controlled or prescribed burning is a technique sometimes used in forest management, farming, or greenhouse gas abatement. Fire is a natural part of both forest and grassland ecology and controlled fire can be a tool for foresters. Controlled burning stimulates the germination of some desirable forest trees, thus renewing the forest.

Fumes from paint, hair spray, varnish, aerosol sprays and other solvents.

Waste deposition in landfills, which generate methane. Methane is highly flammable and may form explosive mixtures with air. Methane is also an asphyxiant and may displace oxygen in an enclosed space. Asphyxia or suffocation may result if the oxygen concentration is reduced to below 19.5% by displacement.

Military resources, such as nuclear weapons, toxic gases, germ warfare and rocketry.

Fertilized farmland may be a major source of nitrogen oxides.

Natural sources

Dust from natural sources, usually large areas of land with little or no vegetation.

Methane, emitted by the digestion of food by animals, for example cattle.

Radon gas from radioactive decay within the Earth's crust. Radon is a colorless, odorless, naturally occurring, radioactive noble gas that is formed from the decay of radium. It is considered to be a health hazard. Radon gas from natural sources can accumulate in buildings, especially in confined areas such as the basement and it is the second most frequent cause of lung cancer, after cigarette smoking.

Smoke and carbon monoxide from wildfires. During periods of active wildfires, smoke from uncontrolled biomass combustion can make up almost 75% of all air pollution by concentration.

Vegetation, in some regions, emits environmentally significant amounts of Volatile organic compounds

(VOCs) on warmer days. These VOCs react with primary anthropogenic pollutants—specifically, NOx, SO_2, and anthropogenic organic carbon compounds — to produce a seasonal haze of secondary pollutants. Black gum, poplar, oak and willow are some examples of vegetation that can produce abundant VOCs. The VOC production from these species result in ozone levels up to eight times higher than the low-impact tree species.

Volcanic activity, which produces sulphur, chlorine, and ash particulates.

"Different Kinds of Pollutants"

Pollutants are classified as primary or secondary. Primary pollutants are usually produced by processes such as ash from a volcanic eruption. Other examples include carbon monoxide gas from motor vehicle exhausts or sulphur dioxide released from factories. Secondary pollutants are not emitted directly. Rather, they form in the air when primary pollutants react or interact. Ground level ozone is a prominent example of secondary pollutants. Some pollutants may be both primary and secondary: they are both emitted directly and formed from other primary pollutants.

Pollutants emitted into the atmosphere by human activity include:

Carbon dioxide – Because of its role as a greenhouse gas it has been described as "the leading pollutant" and "the worst climate pollutant". Carbon dioxide is a natural component of the atmosphere, essential for plant life and given off by the human respiratory system. CO_2 currently forms about 410 parts per million (ppm) of earth's atmosphere, compared to about 280 ppm in pre-industrial times, and billions of metric tons of CO_2 are emitted annually by burning of fossil fuels.

Sulfur oxides (SO_x) – particularly sulphur dioxide, a chemical compound with the formula SO_2. SO_2 is produced by volcanoes and in various industrial processes. Coal and petroleum often contain sulphur compounds, and their combustion generates sulphur dioxide. Further oxidation of SO_2, usually in the presence of a catalyst such as NO_2, forms $H2SO_4$, and thus acid rain. This is one of the causes for concern over the environmental impact of the use of these fuels as power sources.

Nitrogen oxides (NO_x) – Nitrogen oxides, particularly nitrogen dioxide, are expelled from high temperature combustion, and are also produced during thunderstorms by electric discharge. Nitrogen dioxide is a chemical compound with the formula NO_2. It is one of several nitrogen oxides. One of the most prominent air pollutants, this reddish-brown toxic gas has a characteristic sharp, biting odor.

Carbon monoxide (CO) – CO is a colorless, odorless, toxic yet non-irritating gas. It is a product of combustion of fuel such as natural gas, coal or wood. Vehicular exhaust contributes to the majority of carbon monoxide let into our atmosphere. It creates a smog type formation in the air that has been linked to many lung diseases and disruptions to the natural environment and animals.

Volatile organic compounds (VOC) – VOCs are a well-known outdoor air pollutant. They are categorized as either methane (CH_4) or non-methane (NMVOCs). Methane is an extremely efficient greenhouse gas which contributes to enhanced global warming. Other hydrocarbon VOCs are also significant greenhouse gases because of their role in creating ozone and prolonging the life of methane in the atmosphere. This effect varies depending on local air quality. The aromatic NMVOCs benzene, toluene and xylene are suspected carcinogens and may lead to leukemia with prolonged exposure.

Particulate matter / particles, alternatively referred to as particulate matter (PM), atmospheric particulate matter, or fine particles, are tiny particles of solid or liquid suspended in a gas. In contrast, aerosol refers to combined particles and gas. Some particulates occur naturally, originating from volcanoes, dust storms, forest and grassland fires, living vegetation, and sea spray. Human activities, such as the burning of fossil fuels in vehicles, power plants and various industrial processes also generate significant amounts of aerosols. Increased levels of fine particles in the air are linked to health hazards such as heart disease, altered lung function and lung cancer. Particulates are related to respiratory infections and can be particularly harmful to those already suffering from conditions like asthma.

Persistent free radicals connected to airborne fine particles are linked to cardiopulmonary disease.

Toxic metals, such as lead and mercury, especially their compounds.

Chlorofluorocarbons (CFCs) – harmful to the ozone layer; emitted from products are currently banned from use. These are gases which are released from air conditioners, refrigerators, aerosol sprays, etc. On release into the air, CFCs rise to the stratosphere. Here they come in contact with other gases and damage the ozone layer. This allows harmful ultraviolet rays to reach the earth's surface. This can lead to skin cancer, eye disease and can even cause damage to plants.

Ammonia – emitted mainly by agricultural waste. Ammonia is a compound with the formula NH_3. It is normally encountered as a gas with a characteristic pungent odor. Ammonia contributes significantly to the nutritional needs of terrestrial organisms by serving as a precursor to foodstuffs and fertilizers. Ammonia, either directly or indirectly, is also a building block for the synthesis of many pharmaceuticals. Although in wide use, ammonia is both caustic and hazardous. In the atmosphere, ammonia reacts with oxides of nitrogen and sulphur to form secondary particles.

Odors — such as from garbage, sewage, and industrial processes.

Radioactive pollutants – produced by nuclear explosions, nuclear events, war explosives, and natural processes such as the radioactive decay of radon.

Secondary pollutants include:

Particulates created from gaseous primary pollutants and compounds in photochemical smog. Smog is a kind of air pollution. Classic smog results from large amounts of coal burning in an area caused by a mixture of smoke and sulphur dioxide. Modern smog does not usually come from coal but from vehicular and industrial emissions that are acted on in the atmosphere by ultraviolet light from the sun to form secondary pollutants that also combine with the primary emissions to form photochemical smog.

One of the primary components of photochemical smog is Ozone. While ozone in the stratosphere protects earth from harmful UV radiation, ozone on the ground is hazardous to human health. Ground-level ozone is formed when vehicle emissions containing nitrogen oxides (NO_x) and volatile organic compounds (VOCs) from paints, solvents and fuel evaporation interact in the presence of sunlight.

Peroxyacetyl nitrate ($C_2H_3NO_5$) – similarly formed from NO_x and VOCs.

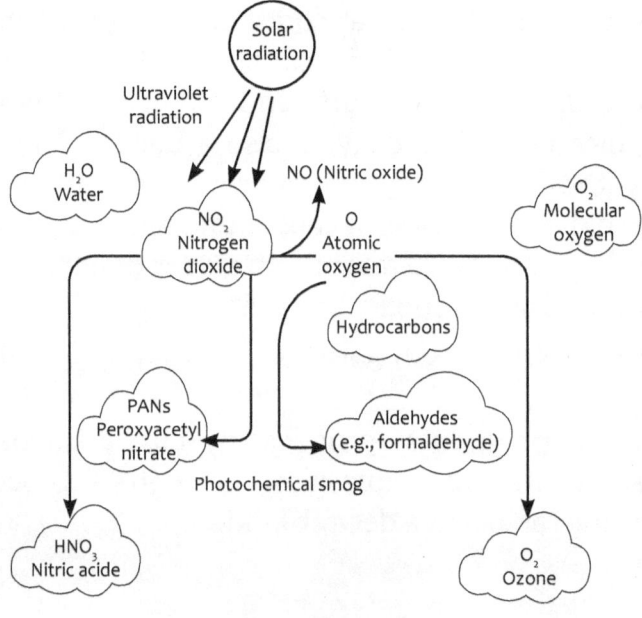

Photochemical Smog

Glossary

1. QSAR—Quantitative structure activity relationship (QSAR) is a strategy of the essential importance for chemistry and pharmacy, based on the idea that when we change a structure of a molecule, then also the activity or property of the substance will be modified.
2. Toxicology—Toxicology is the scientific study of adverse effects that occur in living organisms due to chemicals. It involves observing and reporting symptoms, mechanisms, detection and treatments of toxic substances, in particular relation to the poisoning of humans.
3. Pharmacology—The branch of medicine concerned with the uses, effects, and modes of action of drugs.
4. Biochemistry—Biochemistry is the application of chemistry to the study of biological processes at the cellular and molecular level.
5. Molecular Biology—Molecular Biology is the field of biology that studies the composition, structure and interactions of cellular molecules – such as nucleic acids and proteins – that carry out the biological processes

essential for the cell's functions and maintenance.

6. Acid reflux—a condition in which acidic gastric fluid flows backwards into the oesophagus, resulting in heartburn.

7. Mold culture—A mold or mould is a fungus that grows in the form of multicellular filaments called hyphae. In contrast, fungi that can adopt a single-celled growth habit are called yeasts. That green fungus growing on our bread is called mold. Mold grows where there is damp, decaying organic matter.

8. Tendons—a flexible but inelastic cord of strong fibrous collagen tissue attaching a muscle to a bone.

9. Cartilages—Cartilage is an important structural component of the body. It is a firm tissue but is softer and much more flexible than bone. Cartilage is a connective tissue found in many areas of the body including: Joints between bones e.g. the elbows, knees and ankles and external ear.

10. Nervous system—The network of nerve cells and fibres which transmits nerve impulses between parts of the body. The nervous system consists of the brain, spinal cord, sensory organs, and all of the nerves that connect these organs with the rest of the body. Together, these organs are responsible for the control of the body and communication among its parts.

11. ADHD—ADHD stands for attention deficit hyperactivity disorder. It is a medical condition. A person with ADHD has differences in brain development and brain activity that affect attention, the ability to sit still, and self-control.

12. Allergic rhinitis—Allergic rhinitis, also known as hay

fever, is a type of inflammation in the nose which occurs when the immune system overreacts to allergens in the air. Signs and symptoms include a runny or stuffy nose, sneezing, red, itchy, and watery eyes, and swelling around the eyes.

13. Hives—a rash of round, red welts on the skin that itch intensely, sometimes with dangerous swelling, caused by an allergic reaction, typically to specific foods.

14. Sepsis—a serious condition resulting from the presence of harmful microorganisms in the blood or other tissues and the body's response to their presence, potentially leading to the malfunctioning of various organs, shock, and death.

15. Putrefaction—The process of decay or rotting in a body or other organic matter. In the course of putrefaction, the skin tissues of the body eventually rupture and release the bacterial gas. As the anaerobic bacteria continue consuming, digesting, and excreting the tissue proteins, the body's decomposition progresses to the stage of skeletonization.

16. Mustard gas—a colourless oily liquid whose vapour causes severe irritation and blistering of the skin, used in chemical weapons. Mustard gas, also called sulfur mustard($Cl\text{-}CH_2CH_2)_2S$, gets its name from its sometimes yellow appearance and mustard like smell. It is referred to as a blister agent and comes in vapor, solid or liquid form. In essence, mustard gas kills tissue and membranes in the areas it touches.

17. Adjuvant chemotherapy—Adjuvant treatment is an addition designed to help reach the ultimate goal. Adjuvant therapy for cancer usually refers to surgery followed by chemotherapy or radiotherapy to help

decrease the risk of the cancer recurring (coming back).

18. Radiotherapy—Radiation therapy or radiotherapy, is therapy using ionizing radiation, generally as part of cancer treatment to control or kill cancer cells.

19. Puberty—the period during which adolescents reach sexual maturity and become capable of reproduction.

20. Sudsing—water containing soap or detergent and having bubbles or froth on the surface.

21. Carcinogenicity—Ability of a carcinogen to produce invasive cancer cells from normal cells. A Carcinogen is a substance capable of causing cancer in living tissue.

22. Smoking—Smoking is the process of flavoring, browning, cooking, or preserving food by exposing it to smoke from burning or smoldering material, most often wood. Hot smoking preserves foods in three ways: Heat kills microbes; chemicals found in the smoke — including formaldehyde and alcohols -- act as preservatives; and the food dries out so there is less moist area for bacteria to grow.

23. Rancidity—Rancidity generally is the complete or incomplete oxidation or hydrolysis of fats and oils when exposed to air, light, or moisture or by bacterial action, resulting in unpleasant taste and odor.

24. Fermentation—Fermentation is the process in which a substance breaks down into a simpler substance. Microorganisms like yeast and bacteria usually play a role in the fermentation process, creating beer, wine, bread, yogurt and other foods. Fermentation is an anaerobic process, meaning it doesn't use oxygen.

25. Potency—the power of something to influence or

make an impression. In the field of pharmacology, potency is a measure of drug activity expressed in terms of the amount required to produce an effect of given intensity.

26. Euphoria—Euphoria is an emotional and mental state defined as a sense of great happiness and well-being.
27. DEA—Diethanolamine, DEA is mainly found in moisturizers and sunscreens, while cocamide and lauramide DEA are found in soaps, cleansers, and shampoos.
28. MEA—monoethanolamine or Ethanolamine ETA is an organic chemical compound with the formula $HOCH_2CH_2NH_2$. Ethanolamines are ammonia compounds used in cosmetics as emulsifiers or foaming agents.
29. TEA—Triethanolamine. Triethanolamine (TEA), Diethanolamine (DEA) and Ethanolamine are clear, colorless, thick liquids with ammonia-like odors. In cosmetics and personal care products, Triethanolamine may be used in some makeup products such as eyeliner, mascara, eye shadow, blush, make-up bases and foundations, as well as in fragrances, hair care products, hair dyes, wave sets, shaving products, sunscreens, and skin care and skin cleansing products.
30. Steam distillation—distillation of a liquid in a current of steam, used especially to purify liquids that are not very volatile and are immiscible with water. Steam distillation is the most common way to extract aromatic compounds (essential oil) from a plant. During the steam distillation process, steam passes through the plant material. The combination of heated

steam and gentle pressure causes the essential oil to be released from microscopic protective sacs.

31. Mulching—It is a covering, as of straw, compost, wood chips or plastic sheeting, spread on the ground around plants. This process encourages microbial activity and worms in favor of soil; suppresses growth of weeds; evaporation is reduced; helps to maintain humidity; prevents soil erosion, control weeds and adds nutrients to the soil.

32. Demon TC—It is a pest control product that provides quick knockdown and residual control, as well as a termiticide

33. Cyper TC—Cyper TC is a professional grade insecticide used by those who are licensed to apply termiticide. It is applied to control for wide range of insects, including termites. It can be applied outdoors and is safe for indoor use.

34. Reaction mass—Reaction mass, is a mass against which a system operates in order to produce acceleration. In the case of a rocket, for example, the reaction mass is the fuel shot backwards to provide propulsion.

35. Thrust—The propulsive force of a jet or rocket engine.

36. Oxidizing agent—An oxidizing agent, or oxidant, gains electrons and is reduced in a chemical reaction. Also known as the electron acceptor. An oxidizing agent is needed for the combustion of the fuel in a rocket. The combustion of the fuel generates a lot of energy which pushes out the products of combustion with a very high momentum in the backward direction. This makes it essential to keep the source of oxygen with the fuel.

Glossary

37. Reducing agent—A reducing agent, or reductant, loses electrons and is oxidized in a chemical reaction. Also known as the electron donor.

38. Petrochemicals—Petrochemicals (also known as petroleum distillates) are chemical products obtained from petroleum by refining.

39. Thermoplastic—A thermoplastic, or thermosoftening plastic, is a plastic polymer material that becomes pliable or moldable at a certain elevated temperature and solidifies upon cooling. Examples of thermoplastics are polypropylene, polystyrene, cellulose acetate, PTFE (Teflon), nylon etc.

40. Tenacity—In fiber or textile applications, tenacity is the usual measure of specific strength. Specific strength is also known as the strength-to-weight ratio.

41. E-modulus—An elastic modulus (also known as modulus of elasticity) is a quantity that measures an object or substance's resistance to being deformed elastically (i.e., non-permanently) when a stress is applied to it. A stiffer material will have a higher elastic modulus.

42. Diffusion—Diffusion is the movement of a substance from an area of high concentration to an area of low concentration. Diffusion happens in liquids and gases because their particles move randomly from place to place.

43. Dye baths—A bath prepared for use in dyeing; a solution of coloring matter in which substances to be colored are immersed.

44. Dispersing agent—This agent is used for disperse dyes as these dyes are insoluble in water. By applying this agent, dye molecule breaks and becomes soluble

in the solution.

45. Lignin—A complex organic polymer deposited in the cell walls of many plants, making them rigid and woody. When wood is processed into paper pulp, lignin is produced as a waste product.

46. Hemp—Hemp, or industrial hemp, is a strain of the Cannabis sativa plant species that is grown specifically for the industrial uses of its derived products. Hemp is used to make a variety of commercial and industrial products, including rope, textiles, clothing, shoes, food, paper, bioplastics, insulation, and biofuel.

47. Shellac—Shellac (lac) is a resinous substance prepared from a secretion that encrusts the bodies of a scale insect Karria lacca (Lucifer lacca). Shellac is a natural bioadhesive polymer and is chemically similar to synthetic polymers, and thus can be considered a natural form of plastic.

48. Amber—Amber is the fossilized resin from ancient forests. Large amounts of the dried resin made their way into the sea, sank into deep sediment layers, were covered by dust, sand and stones and were turned into amber in the course of time under high pressure in the absence of air.

49. PVB—Polyvinyl butyral

50. Tensile strength—The resistance of a material to breaking under tension.

51. Cryogenics—Cryogenics is defined as the scientific study of materials and their behaviour at extremely low temperatures.

52. OPCW—Organisation for the prohibition of chemical weapons.

Glossary

53. TNT—TNT is defined as an abbreviation for the explosive trinitrotoluene. An example of TNT is the compound used in dynamite.

54. Heterotrophs—A heterotroph is an organism that cannot manufacture its own food by carbon fixation and therefore derives its intake of nutrition from other sources of organic carbon, mainly plant or animal matter. Herbivores, omnivores, and carnivores: All are examples of heterotroph because they eat other organisms to get proteins and energy.

55. Rancidification—Rancidity generally is the complete or incomplete oxidation or hydrolysis of fats and oils when exposed to air, light, or moisture or by bacterial action, resulting in unpleasant taste and odor.

56. Margarine—Margarine is a spread used for flavoring, baking, and cooking.

57. Ductility—Ductility is the physical property of a material associated with the ability to be hammered thin or stretched into wire without breaking. A ductile substance can be drawn into a wire.

58. Malleability—Malleability is a physical property of metals that defines the ability to be hammered, pressed or rolled into thin sheets without breaking.

59. ICs—Integrated Circuit (IC)

60. Rare—Earth Metals -A rare-earth element or rare-earth metal, as defined by the International Union of Pure and Applied Chemistry, is one of a set of seventeen chemical elements in the periodic table, specifically the fifteen lanthanides, as well as scandium and yttrium.

61. Fatigue life—There are a number of different factors

that can influence fatigue life including the type of material being used, its structure, its shape and temperature changes. In most cases, fatigue life is calculated as the number of stress cycles that an object or material can handle before the failure.

62. Peat—It is a brown deposit resembling soil, also known as turf, is an accumulation of partially decayed vegetation or organic matter.

63. Fissile element—It is a material capable of sustaining a nuclear fission chain reaction. By definition, fissile material can sustain a chain reaction with neutrons of thermal energy.

64. High performance liquid chromatography—High-performance liquid chromatography is a technique in analytical chemistry used to separate, identify, and quantify each component in a mixture. It relies on pumps to pass a pressurized liquid solvent containing the sample mixture through a column filled with a solid adsorbent material.

65. Atomic absorption spectroscopy—AAS is a spectroanalytical procedure for the quantitative determination of chemical elements in which free gaseous atoms absorb electromagnetic radiation at a specific wavelength to produce a measurable signal.

66. Fourier transform infrared spectroscopy—FTIR Spectroscopy is a technique used to obtain an infrared spectrum of absorption or emission of a solid, liquid or gas.

67. Thin layer chromatography—Thin-layer chromatography is a chromatography technique used to separate non-volatile mixtures. Thin-layer chromatography is performed on a sheet of glass,

plastic, or aluminium foil, which is coated with a thin layer of adsorbent material, usually silica gel, aluminium oxide, or cellulose.

68. CSI—Crime Scene Investigation

69. Arson—Arson and explosion investigation is the process of analyzing the charring and chemical residue (if any is left) on the debris found at the crime scene to perform the following tasks: – Determine whether the event was accidental or intentional.

70. TSS—Total suspended solids (TSS) is the dry-weight of suspended particles, that are not dissolved, in a sample of water that can be trapped by a filter that is analyzed using a filtration apparatus. It is a water quality parameter used to assess the quality of a specimen of any type of water or water body, ocean water for example, or wastewater after treatment in a waste water treatment plant.

71. Turbidity—Turbidity is the cloudiness or haziness of a fluid caused by large numbers of individual particles that are generally invisible to the naked eye, similar to smoke in air. The measurement of turbidity is a key test of water quality.

72. Coagulation—Flocculation-In water treatment, coagulation flocculation involves the addition of compounds that promote the clumping of fine particles into larger floc so that they can be more easily separated from the water. The coagulation process involves the addition of the chemical (e.g. alum) and then a rapid mixing to dissolve the chemical and distribute it evenly throughout the water.

73. Sedimentation—Sedimentation is a physical water

treatment process using gravity to remove suspended solids from water. Sedimentation is the process of allowing particles in suspension in water to settle out of the suspension under the effect of gravity. The particles that settle out from the suspension become sediment, and in water treatment is known as sludge.

74. THM precursors—Trihalomethanes (THMs)

75. Ignitability—Ignitability is the characteristic used to define as hazardous those wastes that could cause a fire during transport, storage, or disposal. Examples of ignitable wastes include waste oils and used solvents.

76. Corrosivity—Corrosivity is a measure of how aggressive water is at corroding pipes and fixtures.

77. Reactivity—The rate at which a chemical substance tends to undergo a chemical reaction.

78. Toxicity—the quality of being toxic or poisonous. The toxicity of a drug depends on its dosage. Toxicity refers to how poisonous or harmful a substance can be.

79. DCM—Dichloromethane (DCM), also known as methylene chloride, is a volatile chemical with the formula CH_2Cl_2.

80. Ceviche—a South American dish of marinated raw fish or seafood. Ceviche is marinated in a lime-based mixture. The citric acid from the limes "cooks" the fish so that it can be eaten straight away.

81. Pyrolysis—Pyrolysis is a process of chemically decomposing organic materials at elevated temperatures in the absence of oxygen. The process typically occurs at temperatures above 430°C (800°F) and under pressure. It simultaneously involves the

change of physical phase and chemical composition, and is an irreversible process.

82. Caramelization—Caramelization is the browning of sugar, a process used extensively in cooking for the resulting sweet nutty flavor and brown colour. The brown colours are produced by three groups of polymers: caramelans ($C_{24}H_{36}O_{18}$), caramelens ($C_{36}H_{50}O_{25}$), and caramelins ($C_{125}H_{188}O_{80}$). As the process occurs, volatile chemicals such as diacetyl are released, producing the characteristic caramel flavor.

83. ABS—Acrylonitrile Butadiene Styrene

84. Total alkalinity—Alkaline (basic) substances are present in all water. Within the 7.2-7.8 pH range, alkalinity exists as a bicarbonate material. Total alkalinity, is a measurement of the amount of these substances in parts per million within pools and indicates the water's capacity to withstand changes in pH. The alkaline substances act as buffers, inhibiting changes in pH. Therefore, total alkalinity is a measure of the buffering ability of pool water.

85. Energy orbital—There are multiple orbitals within an atom. Each has its own specific energy level and properties.

86. Emulsifiers—Emulsifiers are molecules with one water-loving (hydrophilic) and one oil-loving (hydrophobic) end. They make it possible for water and oil to become finely dispersed in each other, creating a stable, homogenous, smooth emulsion.

87. Surfactant—a substance which tends to reduce the surface tension of a liquid in which it is dissolved thereby increasing its spreading and wetting

properties. Surface tension is the attractive force exerted upon the surface molecules of a liquid by the molecules beneath that tends to draw the surface molecules into the bulk of the liquid and makes the liquid assume the shape having the least surface area.

88. Viscosity—the state of being thick, sticky, and semi-fluid in consistency, due to internal friction.
89. Gravel—small, rounded stones, often mixed with sand.
90. Silt—fine sand, clay, or other material carried by running water and deposited as a sediment.
91. Aerosols—Aerosols are minute particles suspended in the atmosphere. They are suspension of fine solid particles or liquid droplets, in air or another gas. Aerosols can be natural or anthropogenic.